T0262793

Security and Privacy in Smart Grids

OTHER TELECOMMUNICATIONS BOOKS FROM AUERBACH

Security and Privacy in Smart Grids

Edited by
YANG XIAO

CRC Press
Taylor & Francis Group
Boca Raton London New York

CRC Press is an imprint of the
Taylor & Francis Group, an **informa** business

CRC Press
Taylor & Francis Group
6000 Broken Sound Parkway NW, Suite 300
Boca Raton, FL 33487-2742

First issued in paperback 2018

© 2014 by Taylor & Francis Group, LLC
CRC Press is an imprint of Taylor & Francis Group, an Informa business

No claim to original U.S. Government works

ISBN-13: 978-1-4398-7783-8 (hbk)
ISBN-13: 978-1-138-37462-1 (pbk)

Library of Congress Cataloging-in-Publication Data

Security and privacy in smart grids / editor, Yang Xiao.
    pages cm
    "A CRC title, part of the Taylor & Francis imprint, a member of the Taylor & Francis Group, the academic division of T&F Informa plc."
    Includes bibliographical references and index.
    ISBN 978-1-4398-7783-8 (hardcover : acid-free paper)
    1. Smart power grids--Security measures. I. Xiao, Yang, 1966-

TK3105.S32 2013
621.3190285'58--dc23
                                   2012048623

Visit the Taylor & Francis Web site at
http://www.taylorandfrancis.com

and the CRC Press Web site at
http://www.crcpress.com

Contents

V

Preface

A smart grid is an integration of power delivery systems with communication networks and information technology (IT) to provide better services. Security and privacy will provide significant roles in building future smart grids. The purpose of this edited book is to provide state-of-the-art approaches and novel technologies for security and privacy in smart grids covering a range of topics in these areas.

This book investigates fundamental aspects and applications of smart grids, security, and privacy. It presents a collection of recent advances in these areas contributed by many prominent researchers working on smart grids and related fields around the world. Containing 10 chapters divided into two parts—Part I: Smart Grids in General and Part II: Security and Privacy in Smart Grids, we believe this book will provide a good reference for researchers, practitioners, and students who are interested in the research, development, design, and implementation of smart grid security and privacy.

This work is made possible by the great efforts of our contributors and publisher. We are indebted to our contributors, who have sacrificed days and nights to put together these chapters for our readers. We would like to thank our publisher. Without their encouragement and quality work, we could not have this book.

Yang Xiao
Department of Computer Science
The University of Alabama
Tuscaloosa, Alabama
E-mail: yangxiao@ieee.org

Acknowledgment

This work was supported in part by the U.S. National Science Foundation (NSF) under grants CCF-0829827, CNS-0716211, CNS-0737325, and CNS-1059265.

About the Editor

Dr. Yang Xiao worked in industry as a MAC (Medium Access Control) architect involved in Institute of Electrical and Electronics Engineers (IEEE) 802.11 standard enhancement work before he joined the Department of Computer Science at the University of Memphis in 2002. He is currently a professor in the Department of Computer Science at the University of Alabama. He was a voting member of IEEE 802.11 Working Group from 2001 to 2004. He is an IEEE Senior Member. Dr. Xiao serves as a panelist for the U.S. National Science Foundation (NSF), Canada Foundation for Innovation (CFI) Telecommunications expert committee, and the American Institute of Biological Sciences (AIBS), as well as a referee/reviewer for many national and international funding agencies. His research areas are security, communications/networks, robotics, and telemedicine. He has published more than 200 refereed journal articles and over 200 refereed conference papers and book chapters related to these research areas. Dr. Xiao's research has been supported by the U.S. NSF, U.S. Army Research, the Global Environment for Network Innovations (GENI), Fleet Industrial Supply Center–San Diego (FISCSD), FIATECH, and the University of Alabama's Research Grants Committee. He currently serves as editor in chief for the *International Journal of Security and Networks* (IJSN) and *International Journal of Sensor Networks* (IJSNet). He was the founding editor-in-chief for the *International Journal of Telemedicine and Applications* (IJTA) (2007–2009).

Contributors

Petra Beenken
OFFIS
R&D Division Energy
Oldenburg, Germany

Robert Bleiker
OFFIS
R&D Division Energy
Oldenburg, Germany

Sumit Kumar Bose
Cloud Engineering
Global Technology Center
Unisys Corporation
Oldenburg, Germany

Scott Brock
Cloud Engineering
Global Technology Center
Unisys Corporation
Oldenburg, Germany

Zhe Chen
Tennessee Technological
   University
Cookeville, Tennessee

Xihua Dong
Marvel Semiconductor Inc.
Santa Clara, California

Steffen Fries
Siemens AG
Corporate Technology
Munich, Germany

José González
OFFIS
R&D Division Energy
Oldenburg, Germany

Manimaran Govindarasu
Department of Electrical and
   Computer Engineering
Iowa State University
Ames, Iowa

Nan Guo
Tennessee Technological
 University
Cookeville, Tennessee

Adam Hahn
Department of Electrical and
 Computer Engineering
Iowa State University
Ames, Iowa

Clark Hochgraf
Rochester Institute of
 Technology
Rochester, New York

Hans-Joachim Hof
Department of Computer
 Science and Mathematics
Munich University of Applied
 Sciences
Munich, Germany

Shujie Hou
Tennessee Technological
 University
Cookeville, Tennessee

Rose Qingyang Hu
Department of Electrical and
 Computer Engineering
Utah State University
Logan, Utah

Zhen Hu
Tennessee Technological
 University
Cookeville, Tennessee

Abiodun Iwayemi
Department of Electrical and
 Computer Engineering
Illinois Institute of Technology
Chicago, Illinois

Zhao Li
Industrial Software System
 Group
ABB US Corporation Research
 Center
Raleigh, North Carolina

Chen-Ching Liu
School of Electrical Engineering
 and Computer Science
Washington State University
Pullman, Washington
and
School of Mechanical and
 Materials Engineering
University College Dublin
Dublin, Ireland

Sumita Mishra
Rochester Institute of
 Technology
Rochester, New York

Tae Oh
Rochester Institute of
 Technology
Rochester, New York

Yi Qian
Department of Computer and
 Electronics Engineering
University of Nebraska-Lincoln
Omaha, Nebraska

Robert Qiu
Tennessee Technological
    University
Cookeville, Tennessee

Marbin Pazos-Revilla
Tennessee Technological
    University
Cookeville, Tennessee

Raghuram Ranganathan
Tennessee Technological
    University
Cookeville, Tennessee

Sebastian Rohjans
OFFIS
R&D Division Energy
Oldenburg, Germany

Michael Salsburg
Cloud Engineering
Global Technology Center
Unisys Corporation
Oldenburg, Germany

Ronald Skeoch
Cloud Engineering
Global Technology Center
Unisys Corporation
Oldenburg, Germany

Michael Specht
OFFIS
R&D Division Energy
Oldenburg, Germany

Joern Trefke
OFFIS
R&D Division Energy
Oldenburg, Germany

Mathias Uslar
OFFIS
R&D Division Energy
Oldenburg, Germany

Yongge Wang
Department of Software and
    Information Systems
UNC Charlotte
Charlotte, North Carolina

Zhenyuan Wang
Grid Automation Group
ABB US Corporation Research
    Center
Raleigh, North Carolina

Fang Yang
Grid Automation Group
ABB US Corporation Research
    Center
Raleigh, North Carolina

Yanzhu Ye
Department of Electrical
    Engineering and Computer
    Science
University of Tennessee at
    Knoxville
Knoxville, Tennessee

Peizhong Yi
Department of Electrical and
  Computer Engineering
Illinois Institute of Technology
Chicago, Illinois

Chi Zhou
Department of Electrical and
  Computer Engineering
Illinois Institute of Technology
Chicago, Illinois

PART 1
SMART GRIDS IN GENERAL

1

AN OVERVIEW OF RECOMMENDATIONS FOR A TECHNICAL SMART GRID INFRASTRUCTURE

PETRA BEENKEN, ROBERT BLEIKER, JOSÉ GONZÁLEZ, SEBASTIAN ROHJANS, MICHAEL SPECHT, JOERN TREFKE, AND MATHIAS USLAR

Contents

This chapter introduces the International Electrotechnical Commission Technical Committee (IEC TC) 57 Seamless Integration Architecture (SIA) as a reference architecture for smart grids. It comprises a set of standards that are on various levels essential and widely recommended for smart grid implementations in terms of technical interoperability. Issues like business integration, data definition, applications, field communication for information exchange on the equipment and system interfaces, security, and data management are considered. Each component of the architecture is discussed in detail. As the SIA is not a step-by-step guide to build an information and communications technology (ICT) infrastructure in the energy domain, it is rather a blueprint that focuses on IEC-specific standards. To use the SIA, it is necessary to integrate the architecture in the company workflow or build up an entirely new process. Thus, a methodology is introduced describing how to make the SIA applicable. Finally, further developments of the SIA are listed.

1.1 Introduction

Many national and international smart grid studies, recommendations, and road maps[1-4] have been published recently. Some of them differ in their definition of what the smart grid is and which aspects should be the focus, but all of them agree that standardization is crucial to achieve technical interoperability.

Several standards were identified by most of these studies as core standards (see the work of Rohjans et al.[5,6]). The following standards, which were all developed within the International Electrotechnical Commission Technical Committee (IEC TC) 57, can be regarded as the consensus on essential information technology (IT) standards for the smart grid.

- *IEC 60870: Communication and Transport Protocols[7]*
- *IEC 61334: Distribution Automation[8]*
- *IEC 61400-25: Communication and Monitoring for Wind Power Plants[9]*
- *IEC 61850: Substation Automation Systems and DER* [Distributed Energy Resources][10]
- *IEC 61970/61968: Common Information Model (CIM)[11,12]*
- *IEC 62056: Electricity Metering[13]*
- *IEC 62325: Market Communications Using CIM[14]*
- *IEC 62351: Security for the Smart Grid[15]*
- *IEC 62357: TC 57 Seamless Integration Architecture* [SIA][16]

The TC 57 SIA has a special role as it provides a reference architecture to set the other TC 57 standards in relation to each other and to combine them. It also pursues the objective to identify inconsistencies between the other standards and to resolve them, thus making the whole framework seamless.

This chapter shows the essential standards to reach technical interoperability in a smart grid infrastructure.

1.2 IEC TC 57 Reference Architecture Overview

1.2.1 Introduction to Standardization

In the general scope of smart grids, one has to distinguish between different standardization bodies and other stakeholders for the technical infrastructure to be developed. For the technical infrastructure, most utilities try to adapt to multinational vendors and their corresponding product portfolio. Within this scope, things have changed in the last few years: Whereas typical system committees in standardization had a narrow focus, joint working groups (WGs) have arisen to deal with the bigger picture. User groups have developed to cope with certain aspects like interoperability. The technical base of the smart grid infrastructure now is thoroughly standardized and provides, due to good interoperability checking and tests, many new possibilities for both utilities and vendors. In the very light of international standardization, within the different standardization bodies like ITU (International Telecommunication Union), ISO (International Organization for Standardization), and IEC, the SIA has been identified as the core aspect of future smart grid standardization. Various national road maps like the German, American, and Chinese focus on its aspects and core standards. Furthermore, it is likely to be part of the Korean and Japanese road maps as well. Realizing this, the SIA will be at the very heart of any future standardized smart grid architecture and project.

1.2.2 Main Structure of the Reference Architecture

The IEC Technical Report (TR) 62357 Reference Architecture[16] (Power System Control and Associated Communications—Reference Architecture for Object Models, Services, and Protocols) constitutes a framework for current TC 57 standards. It shows how the various standardization activities within the IEC TC 57 (Power Systems Management and Associated Information Exchange) interrelate and how they contribute to meet the TC's objectives. The reference architecture shows how current standards fit in an overall architecture and provide a seamless integration across systems within the scope of the committee. Aiming to provide a seamless integration, the architecture is also often called the SIA (Seamless Integration Architecture). Like TC 57 addresses business functions in the following domains, these actually comprise the functional scope for the reference architecture:

- Supervisory control and data acquisition (SCADA) and network operation
- Energy management
- Distribution automation
- Customer inquiry
- Meter reading and control
- Substation protection, monitoring, and control

- Records and asset management
- Network expansion planning
- Operational planning and optimization
- Maintenance and construction

Within these domains, the focus of TC 57 is on more abstract data models and generic interfaces at higher levels in the architecture. This comprises an abstract information modeling perspective as well as technology mappings for implementation in all these given areas.

Besides classifying existing standards, areas where harmonization between TC 57 standards is needed and how this could be achieved are identified by the architecture to align and harmonize further standard developments. Ultimately, a future architecture to guide longer-term goals and activities is outlined in IEC TR 62357.

1.2.3 Structure of the Current TC 57 Reference Architecture

Figure 1.1 gives a visual overview of the TC 57 reference architecture as of 2010. The structure of the architecture can be broadly divided into three parts, which are represented by the dashed rectangles A to C in the figure. To structure the various standards and classify their contents, the architecture is partitioned into different layers and pillars (horizontally and vertically). The same shadings indicate the cohesion of standards throughout different layers; in particular, they constitute the pillars in the lower part of the framework. These defined boundaries are finally to depict the coverage of existing standards, allowing identification of harmonization needs. Layers in the first part (A) are mainly concerned with business integration, data definition, and applications, which can be characterized as higher-level abstractions. The first horizontal layer (1) covers standards for integration of different systems and applications (e.g., to business partners or market applications). This could be realized using commercial off-the-shelf middleware in a message-oriented way, as for example often applied in service-oriented architectures (SOAs), in conjunction with the corresponding intersystem/interapplication standards (CIM; eXtensible Markup Language [XML]; CIM Resource Description Framework [RDF]). Standards used on layers 2 and 3 consider the data concepts

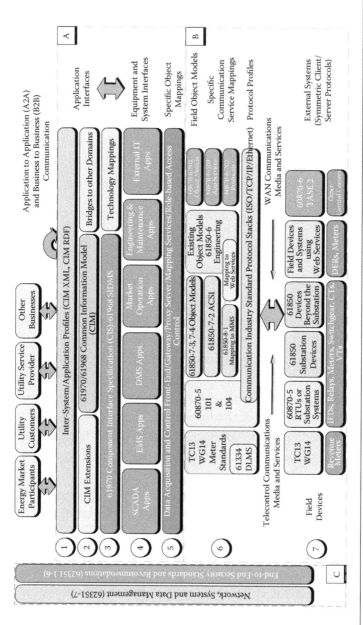

Figure 1.1 Annotated overview of IEC TR 62357 Reference Architecture based on IEC 62357. CT = current transformer; DMS = distribution management system; MMS = manufacturing messaging specification; RTU = remote terminal unit; SIDMS = system interfaces for distribution management systems; TASE = telecommunication application service element 2; TCP/IP = Transmission Control Protocol/Internet Protocol; VT = voltage transformer; WAN = wide-area network; WG = Working Group. (Reprinted with permission from International Electrochemical Commission. *IEC 62357, 2nd edition: TC 57 Architecture—Part 1: Reference Architecture for TC 57—Draft*, 2009. Geneva, Switzerland: IEC.)

and interfaces for the focused applications (layer 4). These applications serve as central IT-driven elements for power systems control and operations. There are two aspects to consider for these applications: the upper integration using corresponding interfaces (application interfaces) and the lower integration (equipment and system interfaces). To allow for successful integration, the systems must be enabled to be supplied with operation-relevant data (e.g., from technical devices like substations) and further provide other IT systems and applications with important data. Currently, gathering data and controlling field devices require data and communication mappings between different standards due to a variety of access options and data formats. For these cases, abstractions to encapsulate access to the required technical information are offered by layer 5, namely, the SCADA front end.

Below this layer, the architecture is structured in four pillars containing mainly standards dealing with more technical field communication for information exchange on the equipment and system interfaces (part B). Each pillar addresses standards for different device categories: revenue meters; intelligent electronic devices (IEDs), relays, meters, switchgear, current transformers (CTs), voltage transformers (VTs); distributed energy resources (DER), meters; and other control centers.

The upper layers of this part (6) include standards containing object models for field devices and device components, specific communication service mappings, and protocol profiles. At this point, communication to exchange data usually takes place through wide-area networks (WANs) of geographically separated locations using standard protocol stacks like the ISO Open System Interconnection (OSI) model or the Internet Protocol stack using the Transmission Control Protocol/Internet Protocol (TCP/IP) and Ethernet. Standards for the different devices and systems to communicate with are finally depicted in layer 7.

Vertical layers on the left (C) indicate cross-cutting standards that especially focus on security and data management addressed by the IEC 62351 standards family. In these standards, each horizontal layer is addressed by individual parts to meet specific requirements. As these vertical layers span the whole framework, they can be considered a highly important factor for successful integration, and in the end, they contribute to secure systems operation.

1.2.4 Future Vision of a Seamless Integration

Based on the findings from reviewing the current reference architecture, the need for a long-term architecture vision was determined, going further than just harmonization between different standards. As a start, the committee agreed on 16 architectural principles, for instance, about the focus of the ongoing work, harmonization efforts for existing standards, and the definition of criteria to ensure a system's compliance to the reference architecture. Starting with these principles, a strategy adopting the CIM and other abstract information models as the source of the semantics as basis for future standards development is presented. This may lead to reduced execution times and can potentially avoid information loss due to the mapping of different language concepts on different layers, which can finally ease integration.

In the following sections, a closer look at the standards and the different aspects, aligned with the different parts of the current reference architecture, is provided. These sections are "Integration of Business Partners and Applications" (Section 1.2.5), "Integration of Energy Systems" (Section 1.2.6), and "Security and Data Management" (Section 1.2.7).

1.2.5 Integration of Business Partners and Applications

The top part of the SIA as illustrated in Figure 1.2 addresses the integration of business partners, Business to Business (B2B), and applications, Application to Application (A2A). Key elements of this part are therefore market participants like utility customers, utility service

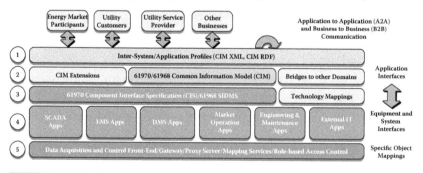

Figure 1.2   Top part of the SIA.

providers, or other business participants and IT applications within utility companies like SCADA or EMS (energy management systems).

The top part of the SIA can be divided into five layers: market communication (1), core data model (2), integration of applications (3), applications (4), and equipment and system interface (5). Layers 1–5 are described next:

- Layer 1 covers the integration of market participants and their IT systems based on the IEC CIM and its serialization in different formats like XML or RDF. In addition, the IEC 62325 series describes the use of the CIM for market communications between business partners. Communication is described independent of technology but relying on interapplication messaging as provided by commercial off-the-shelf middleware.
- Layer 2 provides the IEC 61970-301 and 61968-11 standards, which describe the CIM data model. The CIM is the core data model within the SIA for usage within data exchange addressing both types of integration, B2B and A2A. The CIM is a data model for abstract and physical objects in the electricity domain. As requirements change and each utility is different, custom extensions of the CIM might be necessary (CIM extensions). In particular, these extensions will become necessary when dealing with data not strictly belonging to the electricity domain (bridges to other domains).
- Layer 3 focuses on integration of transmission and distribution IT applications. On the one hand, IEC 61970-401 provides application interfaces for EMSs. On the other hand, the IEC 61968 standards series describes an Enterprise Application Integration (EAI) framework for exchanging data between distribution management systems (DMSs). In the course of new technologies, technology mappings might be necessary.
- Layer 4 shows various transmission and distribution IT components of a utility application landscape. This includes the following systems:
  - SCADA: Real-time system that supports the control room operation, including data acquisition and supervisory control using remote terminal units (RTUs) in the substations.[11]

- EMS: Computer system providing basic services and a set of applications to support the effective operation of electrical generation and transmission facilities.[17] Within this, monitor and control functionality is provided by SCADA systems.
- DMS: Several distributed application components supporting the management of electrical distribution networks.[11] These components provide capabilities like monitoring and control of equipment for power delivery, management processes to ensure system reliability, voltage management, demand-side management, outage management, and work management.
- Market operations applications: Dealing with data exchange between market participants, supporting processes like customer switching or meter data exchange.
- Engineering and maintenance applications: Supporting processes like network maintenance and extension planning.
- External IT applications: Applications that are not strictly utility systems like customer resource management systems.[16]

- Layer 5 addresses the integration of IT systems of layer 4 and external systems and technical devices in the field. Therefore, this layer describes an equipment and system interface to acquire data or control devices. Applications listed in layer 4 act as clients that connect to remote servers in the field, whereas the connection can be established through various communication networks and technologies. Layer 5 is the last layer of the top part of the SIA and connects the top part of the SIA with the lower part (see the dashed rectangles A and B in Figure 1.1).

Standards listed in this part of the SIA are all developed within WGs of IEC TC 57, *Power Systems Management and Associated Information Exchange.*

In the following, the core standards series of the upper part of the SIA (IEC 61970, IEC 61968, and IEC 62325) as well as their essential contributions, the IEC CIM, the Component Interface Specification (CIS), and the IEC Interface Reference Model (IRM), are introduced.

1.2.5.1 IEC 61970: Energy Management System Application Program Interface The IEC 61970 standards series defines application program interfaces (APIs) for EMS to support the integration of applications developed by different suppliers in the control center environment and the exchange of information to systems external to the control center environment.[12] An overview of the EMS APIs is provided in Figure 1.3. The following parts of IEC 61970 are currently available:[18]

- *IEC 61970-1 Ed. 1.0: Guidelines and General Requirements*
- *IEC/TS 61970-2 Ed. 1.0: Glossary*
- *IEC 61970-301 Ed. 2.0: Common Information Model (CIM) Base*
- *IEC/TS 61970-401 Ed. 1.0: Component Interface Specification (CIS) Framework*
- *IEC 61970-402 Ed. 1.0: Common Services*
- *IEC 61970-403 Ed. 1.0: Generic Data Access*
- *IEC 61970-404 Ed. 1.0: High Speed Data Access (HSDA)*
- *IEC 61970-405 Ed. 1.0: Generic Eventing and Subscription (GES)*
- *IEC 61970-407 Ed. 1.0: Time Series Data Access (TSDA)*
- *IEC 61970-453 Ed. 1.0: CIM Based Graphics Exchange*
- *IEC 61970-501 Ed. 1.0: Common Information Model Resource Description Framework (CIM RDF) Schema*

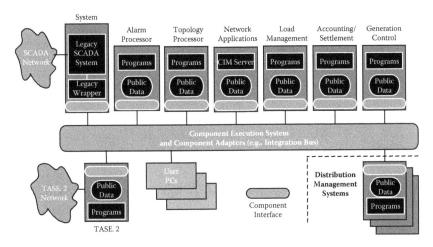

Figure 1.3 Overview of the EMS-API. PC, personal computer. (Reprinted with permission from International Electrochemical Commission. *61968-1: Application Integration at Electric Utilities— System Interfaces for Distribution Management Part 1: Interface Architecture and General Requirements*, 2007. Geneva, Switzerland: IEC.)

The IEC TC 57 WG 13 EMS API is in charge of the development of the IEC 61970 series. The IEC 61970 series, in particular the CIM, is unanimously recommended for smart grid architectures.

1.2.5.2 IEC 61968: Application Integration at Electric Utilities—System Interfaces for Distribution Management The IEC 61968 standards series aims at facilitating the interapplication integration of the various distributed software application systems supporting the management of utility's electrical distribution networks.[11] In contrast to the general understanding of interapplication integration, focusing on programs in the same application system, the IEC 61968 series aims at integrating disparate loosely coupled applications within utility enterprises that are already built or new (legacy or purchased applications). Here, connections between applications are established via middleware services that broker messages. IEC 61968 has the following parts:[18]

- *IEC 61968-1 Ed. 1.0: Interface Architecture and General Requirements*
- *IEC/TS 61968-2 Ed. 1.0: Glossary*
- *IEC 61968-3 Ed. 1.0: Interface for Network Operations*
- *IEC 61968-4 Ed. 1.0: Interfaces for Records and Asset Management*
- *IEC 61968-9 Ed. 1.0: Interfaces for Meter Reading and Control*
- *IEC 61968-11 Ed. 1.0: Common Information Model (CIM) Extensions for Distribution*
- *IEC 61968-13 Ed. 1.0: CIM RDF Model Exchange Format for Distribution*

IEC TC 57 WG 14: System Interfaces for Distribution Management (SIDM) is responsible for the development of the IEC 61968 series.

1.2.5.3 IEC 62325: Framework for Energy Market Communications The IEC 62325 aims at describing the use of the CIM for market communications between business partners. The term *market communications* refers to data exchange between market participants like energy suppliers or distribution system operators along the electricity value chain. Here, WG 16 of the IEC TC 57 develops a framework for communications in a deregulated electricity market. The IEC 62325 consists of the following parts:[18]

- *IEC/TR 62325-101 Ed. 1.0: General Guidelines*
- *IEC/TR 62325-102 Ed. 1.0: Energy Market Model Example*
- *IEC/TR 62325-501 Ed. 1.0: General Guidelines for Use of Electronic Business Using XML (ebXML)*
- *IEC/TS 62325-502 Ed. 1.0: Profile of ebXML*

The IEC 62325 series is being developed by the IEC TC 57 WG 16 (*Deregulated Energy Market Communications*). In contrast to the IEC 61968 and 61970 standards series, this series still contains many parts that are still the subject of future work (see IEC 62325-101).[14]

As communication between market participants in the electricity domain is subject to national regulation, application of these standards requires analysis of current national regulations, laws, and guidelines. National guidelines may force the application of specific data formats and protocols not considered within IEC 62325. In Germany, for instance, the Electronic Data Interchange for Administration, Commerce, and Transport (EDIFACT) format is currently required for data exchange between market participants for processes like customer switching.

1.2.5.4 The IEC Common Information Model   The IEC CIM is a very large abstract data model describing abstract (like documents) as well as physical (like power transformer) objects of the energy domain. It was originally created to solve the problem of vendor lock-in by EMS.[19] Many aspects of the power system of concern to TC 57 are modeled only using the CIM, like generation equipment or energy schedules.[16] However, other parts are modeled in both the CIM and in the IEC 61850 standards developed by WG 10 (e.g., substation equipment, including transformers, switches, or breakers).[16]

The idea of the CIM was to provide a common information model that should support the exchange of information between different EMS components and thus enable the interconnection of applications from different vendors. The CIM was originally developed within several projects sponsored by the Electric Power Research Institute (EPRI). Over time, the CIM was extended to fit the needs of distribution management; at the moment, WG 16 is extending the CIM for use within market communication. Currently, TC 57 WG 13, WG 14, WG 16, and WG 19 are involved in the development of the

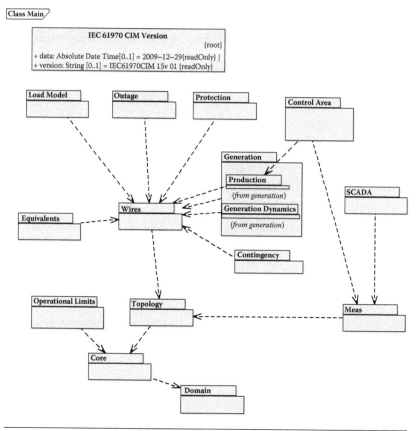

Figure 1.4   Overview of the IEC CIM.

CIM.[20] Furthermore, many members of the WGs joined the work of the CIM Users Group (CIMug; http://cimug.ucaiug.org).

The formal definition of the CIM is done using the Unified Modeling Language (UML); an overview is depicted in Figure 1.4. The model includes public classes and attributes describing (real and abstract) objects of the energy domain as well as relationships between them. It is currently maintained in the SparxSystems Enterprise Architect. For better maintenance, the various classes are grouped in corresponding packages, and the different WGs focus on different packages and describe them in different parts of the standards series, basically IEC 61968-11 and 61970-301.

Whereas the standards documents related to the CIM are developed within the IEC, the electronic UML model is hosted at the

CIMug site. Therefore, CIMug members have access to the model without the need to participate in the IEC standardization process.

It is difficult and often not necessary to use the whole model within a project or company. To make the use of the CIM more applicable, profiles of the CIM that only include essential classes and associations of the CIM are used. On the one hand, single companies use intracorporate profiles; on the other hand, large profiles exist that are partly standardized and widespread within the utility domain:

- **CPSM**: The Common Power System Model (CPSM) is used in the United States for the exchange of transmission system models.[21]
- **CDPSM**: The Common Distribution Power System Model (CDPSM) is used in Europe for the exchange of distribution power system models.[22]
- **ENTSO-E**: The European Network of Transmission System Operators for Electricity (ENTSO-E; http://www.entsoe.eu/) profile is used in Europe for the exchange of transmission system models.
- **ERCOT**: The Electric Reliability Council of Texas (ERCOT; http://www.ercot.com/) profile is an intracorporate data model.

The main application scenarios for the CIM are as follows:[23]

- **Exchange of topology data:** Supporting the exchange of power system models between systems through CIM profiles for transmission (CPSM) and distribution (CDPSM) networks. In addition, a corresponding serialization of these profiles for XML and RDF is defined in the standard series IEC 61968 (distribution) and IEC 61970 (transmission). This enables standards-based exchange of static and dynamic data as well as the current state of electrical networks.
- **Coupling of applications:** Using standard-based interfaces as described in the standard documents IEC 61968 Part 3-9 and IEC 61970-4xx. Here, the CIM provides the semantics for the underlying data of the specified interfaces. This supports integration of applications of different vendors within application landscapes in utilities.

- **XML-based message exchange with CIM semantics:** Can be used to build personal XML schemas to enable standards-based message exchange between applications. As with coupling of applications, the CIM provides a standardized semantics for coupling applications of different vendors. A tool for developing such schemas is available, for example through Langdale Consultants (http://www.cimtool.org).

In the following, some characteristics of the CIM are summarized:[16]

- **The CIM is hierarchical:** Common classes inherit common attributes to subclasses.
- **The CIM is normalized:** All attributes are unique and belong to only one class. The use of attributes within other classes is done by defining relationships between these classes. Relationships supported include generalization, association, and aggregation.
- **The CIM addresses the static (or structural) model view:** In the CIM, physical objects may be represented by several interrelated classes. The objects one application may want to access are not grouped in a single class. Therefore, the model is not appropriate for adding dynamics in the form of operations or methods to the actual class definitions.
- **The CIM is modeled in UML:** The entire CIM is provided as a UML model file.
- **The CIM UML model is the basis for the standards:** The corresponding IEC standards documents are autogenerated using the electronic UML model.
- **The CIM has a representation in XML:** See the described CIM application scenarios, like exchange of topology data using CPSM and CDPSM or XML-based message exchange.
- **The CIM is in use in many production systems:** For example, in the United States the use of the CIM for data exchange is prescribed in several states. In Europe, the CIM is used for the exchange of transmission system models by European Transmission system operators organized in the ENTSO-E.
- The CIM is meant to contain classes and attributes that will be exchanged over public interfaces between major applications.

The maintenance process is continuously improving the model using the UML format. Once a year, a new release is published; the current release is version 15. Proposals for the extension or amendment of the CIM are done via the CIMug site. Here, CIMug members can enter modeling issues that will be discussed later in modeling team meetings and may finally lead to changes of the CIM.

1.2.5.5 Component Interface Specification   The IEC 61970-4xx standards documents basically provide CIS and Generic Interface Definitions (GIDs) that define interfaces and APIs for a standards-based integration of applications or components of EMS. The purpose of the CIS is to specify the interfaces that an application or system should use to facilitate message-based integration with other independently developed applications or systems.[16] On the one hand, the CIS specifies the information content of the messages; on the other hand, it defines what services should be used to convey the messages. This way, a clear definition of what and how information is available for processing and expected by receiving applications is provided. Furthermore, the CIS enables a single adapter to be built for a given infrastructure technology independent of who developed the other systems.

Since multiple application categories require many component interface services, the service definitions are specified as generic services independent of the particular application that uses them.[16] The GID is the collection of these generic services. Due to the many generic services the IEC 61970-4xx standards series comprises, the following subparts consider the various types of data exchange:[16,23]

- **IEC 61970-401 CIS framework:** Describes scope and vision of the CIS.
- **IEC 61970-402 CIS—common services:** Describes common services that serve as basis for the GID. Here, the CIM semantic is used for data definitions in interfaces.
- **IEC 61970-403 CIS—generic data access:** Defines interfaces that can be used to read and write real-time data. These interfaces provide a request/reply-oriented service for access of complex data structures.
- **IEC 61970-404 HSDA:** Describes interfaces that can be used for high-performance access of simple data structures.

- **IEC 61970-405 GES:** Defines interfaces that can be used to monitor events and alarms based on publish and subscribe methods.
- **IEC 61970-407 TSDA:** Describes interfaces that can be used to access aggregated historical data.

Currently, the replacement of the aforementioned IEC 61970-403 and -407 standards is planned by the IEC. Instead of these standards, the corresponding standards of the OPC Unified Architecture (UA) shall be used in the future.

Implementing a specific type of application requires defining what object classes and attributes are exchanged as well as what interface is used.[16] These object classes and attributes typically consist of subsets or views of the CIM object classes. In conclusion, the CIM data model defines "which" data can be exchanged; the CIS and GID specifies "how" these data can be exchanged.[20]

In addition, following the Open Management Group (OMG) Model Driven Architecture (MDA) approach[24] descriptions based on the concepts of the platform-independent model* (PIM) and the platform-specific model† (PSM) are provided. First, the Part 4xx series of the 61970 standards provides the PIM component models of the CIS, defining interfaces in terms of events, methods, and properties independent of the underlying infrastructure.[16] Second, the Part 5xx series of the 61970 standards defines the technology mappings to technologies such as C++, Java, Web Services, and XML.[16]

1.2.5.6 The Interface Reference Model The IRM illustrated in Figure 1.5 and described in the IEC 61968-1 standard, Interface Architecture and General Recommendations[11] defines interfaces for the major components of a DMS. The purpose of the IRM and the individual system interfaces defined therein is to provide a framework for a series of message payload standards based on the CIM. These message payload standards are the subject of the IEC 61968-3 to –9

---

* A platform-independent model is a view of a system from the platform-independent viewpoint.[24]

† A platform-specific model is a view of a system from the platform-specific viewpoint.[24]

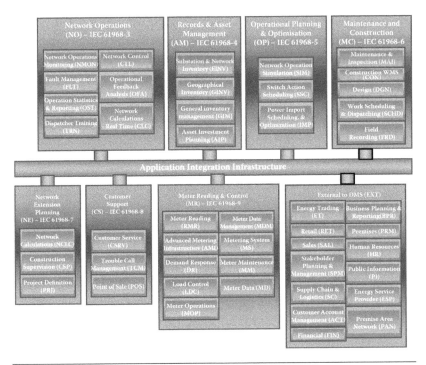

Figure 1.5 Overview of the IEC 61968 IRM. (Reprinted with permission from International Electrotechnical Commission (IEC). *61968-1: Application Integration at Electric Utilities—System Interfaces for Distribution Management Part 1: Interface Architecture and General Requirements (Draft)* (2010). Geneva, Switzerland: IEC.)

standards.[16] The IRM aims at supporting interoperability between these components independent of systems, platforms, and languages.

Within the IEC 61968-3 to −9 standards, the use of XML for the exchange of information between the various systems is specified.[16] Here, several use cases are provided that define the data content of message payloads between these various systems. Furthermore, XML schemas are used to define the structure and format for each message payload. The message payloads defined here are intended to be leveraged by both service-oriented architectures (SOAs) and enterprise service buses (ESBs). In the future, it is possible that payload formats other than XML could also be adopted.[16] The IRM illustrates seven domains supporting core business functions of distribution management. Each domain contains several abstract components and shows the relevant IEC 61968 part (-3 to −9) where interface definitions for these components are described. In addition, components external but related to DMS are grouped in their own domain external to

DMS (EXT). All components are integrated through a CIM-based, message-oriented middleware (MOM)—the application integration infrastructure. The application integration infrastructure acts here as an enabler for XML-based message exchange with CIM semantics as described in Section 1.2.5.4.

Figure 1.5 shows only the top-level business functions and business subfunctions of the IRM. A detailed, table-based description, containing the following elements, is provided in the IEC 61968-1 standard:[25]

- **Business functions:** Like network operations or records and asset management; see Figure 1.5.
- **Business subfunctions:** Like network operations monitoring or substation and network inventory; see Figure 1.5.
- **Abstract components:** Are grouped by business subfunctions and define abstract logical components like SCADA simulation or substation state supervision. It is expected that concrete physical applications of vendors will provide the functionality of one or more abstract components.[11]

After having explained the upper business integration part of the SIA in this section, the following section is about the integration of energy systems that deals with the connection to information exchange on the equipment and system interfaces.

1.2.6 Integration of Energy Systems

The lower part (part B in Figure 1.1) of the SIA, shown in Figure 1.6, can be divided into four layered pillars. The basement of each pillar is a

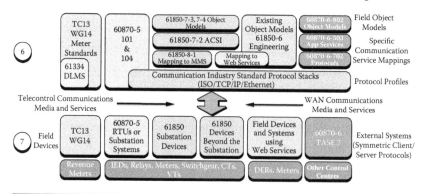

Figure 1.6   Lower part of the SIA.

Figure 1.7 Pillar for revenue meters.

group of different field devices (revenue meters, Section 1.2.6.1; IEDs, relays, meters, switchgear, CT, and VT in Section 1.2.6.2; DER and meter, see Section 1.2.6.3; other control centers, Section 1.2.6.4). The next layer (7) describes external communication systems for the field devices, which are connected to the following layers, including protocol profiles, specific communication service mappings, and field object models. The top of the pillars (6) is linked to the SCADA front-end layer of the SIA.

1.2.6.1 Revenue Meters   The first pillar (see Figure 1.7) includes the communication of revenue meters, which is based on standards from the IEC TC 13 WG 14. Among others, the standard series IEC 61334 is mentioned. Revenue meters include the various types of smart meters for residential, commercial, and industrial billing.

1.2.6.1.1 TC 13 WG 14   The IEC TC 13 WG 14 name is *Data Exchange for Meter Reading, Tariff and Load Control*. Its task is to establish standards, by reference to ISO/OSI standards, necessary for data exchanges by different communication media, for automatic meter reading, tariff and load control, and consumer information. Thereby, the media can be distribution line carrier (DLC), telephone (including Integrated Services Digital Network [ISDN]), radio, or other electrical or optical system, and they may be used for local or remote data exchange. Furthermore, they are acting in category D liaison with the DLMS (Distribution Line Message Specification) User Association (UA; http://www.dlms.com/index2.php).

The TC 13 Strategic Business Plan (http://www.iec.ch/cgi-bin/ getfile.pl/sbp_13.pdf?dir=sbp&format=pdf&type=&file=13.pdf) from 2009 specifies future activities of the WG 14. One main objective is to focus on the extension of the IEC 62056 to support smart metering, which includes the extension of the COSEM data model. The model has to deal with new functions and new DLMS-based messaging methods as well as communication profiles have to be added. Furthermore, standards from other TCs shall be used whenever it is appropriate and close cooperation with the DLMS UA and industry consortia are planned.

1.2.6.1.2 TC 13 WG 14 Meter Standards   The TC 13 WG 14 mainly deals with the development of the metering standards IEC 62056 and IEC 62051, which are presented in the following:

- IEC 62056: *IEC 62056, Electricity Metering—Data Exchange for Meter Reading, Tariff, and Load Control,* consists of several substandards dealing with DLMS and COSEM. The following six parts comprise the main specifications:[18]
  - *IEC 62056-21 Ed. 1.0: Direct Local Data Exchange*
  - *IEC 62056-42 Ed. 1.0: Physical Layer Services and Procedures for Connection-Oriented Asynchronous Data Exchange*
  - *IEC 62056-46 Ed. 1.1: Data Link Layer Using HDLC* [High-Level Data Link Control] *Protocol*
  - *IEC 62056-53 Ed. 2.0: COSEM Application Layer*
  - *IEC 62056-61 Ed. 2.0: Object Identification System (OBIS)*
  - *IEC 62056-62 Ed. 2.0: Interface Classes*

  In part 21,[13] an alternate protocol stack is defined that is based on ASCII. MODE E is introduced as a new mode enabling negotiations to a switchover to COSEM/HDLC— defined in parts 46[26] and 53[27]—for clients. As a result of the switchover, following communications will be based on the COSEM/HDLC protocol stack. HDLC defines a standard data link layer, ensuring a reliable transport of COSEM data packages in a client-server architecture. Thereby, the layer performs functions like low-level addressing, data integrity checks, data sequencing, and segmentation as well as assimilation, link-level handshaking, and data flow control.

COSEM specifies a protocol for application layers that covers basic functionalities like set, get, and action operations within the meters. Beyond these basic functions, COSEM also allows handling of access rights and client-server connections, abstracting meter data from/to COSEM class instances, framing data into COSEM packages, and high-level segmentation of data into blocks.

Physical layer services needed for the data communication are specified in part 42.[28] Part 61[29] includes the OBIS, which defines a standard list of meter data object identifiers. Those identifiers are defined as six-character codes for each object, and they are maintained by the DLMS UA. Part 62[30], as the last main part of the series, considers standard interface classes. They can be used to represent all possible meter data, which are abstracted into high-level objects. Finally, the protocol stacks can operate on the high-level objects.

In addition, the standard series includes the following parts:[18]

- *IEC 62056-31 Ed. 1.0: Use of Local Area Networks (LANs) on Twisted Pair with Carrier Signaling*
- *IEC/TS 62056-41 Ed. 1.0: Data Exchange Using Wide Area Networks: Public Switched Telephone Network (PSTN) with LINK+ Protocol*
- *IEC 62056-47 Ed. 1.0: COSEM Transport Layers for IPv4* [Internet Protocol Version 4] *Networks*
- *IEC/TS 62056-51 Ed. 1.0: Application Layer Protocols*
- *IEC/TS 62056-52 Ed. 1.0: Communication Protocols Management Distribution Line Message Specification (DLMS) Server*
- IEC/TR 62051: The second standard series maintained by WG 14 is *IEC 62051*[31] *Electricity Metering—Data Exchange for Meter Reading, Tariff, and Load Control*, which is a relatively short series. It provides definitions of specific terms used for drafting standards within the context of electrical measurement, tariff, and load control as well as customer/utility information exchange systems. The set of provided definitions is completed by those terms already dealt with in IEC 60050 (http://www.electropedia.org/). The defined terms could also

be used for upcoming standards coping with electricity pre-payment systems and the dependability of electricity metering equipment.

- *IEC/TR 62051 Ed. 1.0: Glossary of Terms*
- *IEC/TR 62051-1 Ed. 1.0: Terms Related to Data Exchange with Metering Equipment Using DLMS/COSEM*

1.2.6.1.3 IEC 61334 DLMS   The TC 57 WG 9 develops the standard series *IEC 61334, Distribution Automation Using Distribution Line Carrier Systems.* Those standards are mainly focusing protocols used to enable the communication from the distribution control center to distribution automation field devices using the distribution grid. The application area of the standards series contains the communication by carrier systems on the middle-voltage layer as well as on the low-voltage layer. Thereby, the DLC systems enable a bidirectional communication for various devices and functions like control centers, data concentrators, load management, or streetlights.

Based on a client-server architecture, the substandard IEC 61334-4-1,[32] which is also known as the DLMS, defines a reference architecture and provides an abstract and object-oriented server model. The server model explicitly takes limited hardware resources and the low bandwidth of distribution equipment into consideration. Abstract Syntax Notation One (ASN.1) is used to describe the protocol data units (PDUs) of the application protocol of the model. IEC 61334-6[33] adds efficient coding possibilities to this description. The substandards IEC 61334-5-1 to –5-5[34–38] define different physical and Media Access Control layers with different modulation technologies that are applicable for both low- and medium-voltage grids. IEC 61334-4-511[39] and –4-512[40] define a management framework and techniques that are especially aligned to IEC 61334-5-1. IEC 61334-3-21[41] and –3-22[42] define requirements to feed DLC signals into middle-voltage lines without violating security issues.

Currently, the standard series includes the following parts:[18]

- *IEC/TR 61334-1-1 Ed. 1.0: General Considerations— Distribution Automation System Architecture*
- *IEC/TR 61334-1-2 Ed. 1.0: General Considerations—Guide for Specification*

- *IEC/TR 61334-1-4 Ed. 1.0: General Considerations—Identification of Data Transmission Parameters Concerning Medium- and Low-Voltage Distribution Mains*
- *IEC 61334-3-1 Ed. 1.0: Mains Signaling Requirements—Frequency Bands and Output Levels*
- *IEC 61334-3-21 Ed. 1.0: Mains Signaling Requirements—MV Phase-to-Phase Isolated Capacitive Coupling Device*
- *IEC 61334-3-22 Ed. 1.0: Mains Signaling Requirements—MV Phase-to-Earth and Screen-to-Earth Intrusive Coupling Devices*
- *IEC 61334-4-1 Ed. 1.0: Data Communication Protocols—Reference Model of the Communication System*
- *IEC 61334-4-32 Ed. 1.0: Data Communication Protocols—Data Link Layer—Logical Link Control (LLC)*
- *IEC 61334-4-33 Ed. 1.0: Data Communication Protocols—Data Link Layer—Connection Oriented Protocol*
- *IEC 61334-4-41 Ed. 1.0: Data Communication Protocols—Application Protocol—Distribution Line Message Specification*
- *IEC 61334-4-42 Ed. 1.0: Data Communication Protocols—Application Protocols—Application Layer*
- *IEC 61334-4-61 Ed. 1.0: Data Communication Protocols—Network Layer—Connectionless Protocol*
- *IEC 61334-4-511 Ed. 1.0: Data Communication Protocols—Systems Management—CIASE Protocol*
- *IEC 61334-4-512 Ed. 1.0: Data Communication Protocols—System Management Using Profile 61334-5-1—Management Information Base (MIB)*
- *IEC 61334-5-1 Ed. 2.0: Lower-Layer Profiles—The Spread Frequency Shift Keying (S-FSK) Profile*
- *IEC/TS 61334-5-2 Ed. 1.0: Lower-Layer Profiles—Frequency Shift Keying (FSK) Profile*
- *IEC/TS 61334-5-3 Ed. 1.0: Lower-Layer Profiles—Spread Spectrum Adaptive Wideband (SS-AW) Profile*
- *IEC/TS 61334-5-4 Ed. 1.0: Lower-Layer Profiles—Multicarrier Modulation (MCM) Profile*
- *IEC/TS 61334-5-5 Ed. 1.0: Lower-Layer Profiles—Spread Spectrum–Fast Frequency Hopping (SS-FFH) Profile*
- *IEC 61334-6 Ed. 1.0: A-XDR Encoding Rule*

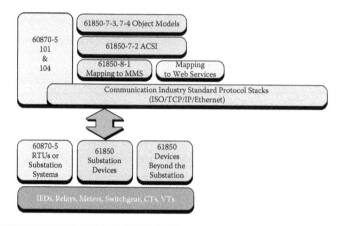

Figure 1.8   Pillar for IED, relays, meters, switchgear, CT, and VT.

1.2.6.2 IEDs, Relays, Meters, Switchgear, CTs, and VTs   The second pillar (see Figure 1.8) covers the monitoring and control of IEDs, common relays, meters, and switchgears as well as CTs and VTs. This large group of field devices mainly uses communication standardized by the two standard series IEC 61850 and IEC 60870-5, including their substandards.

1.2.6.2.1 IEC 60870-5 RTUs or Substation Systems   The development of the standard series *IEC 60870-5, Telecontrol Equipment and Systems—Part 5: Transmission Protocols*, was started in the 1980s by TC 57 WG 3. The main objective was to develop an internationally standardized communication protocol for telecontrol applications in distributed power networks. In the early 1990s, the first five standards were published:[18]

- *IEC 60870-5-1 Ed. 1.0: Transmission Frame Formats*
- *IEC 60870-5-2 Ed. 1.0: Link Transmission Procedures*
- *IEC 60870-5-3 Ed. 1.0: General Structure of Application Data*
- *IEC 60870-5-4 Ed. 1.0: Definition and Coding of Application Information Elements*
- *IEC 60870-5-5 Ed. 1.0: Basic Application Functions*

Then, all further standards dealing with special applications should be published as companion standards. To this date, the following four companion standards have been published and are widely used:[18]

- *IEC 60870-5-101 Ed. 2.0: Transmission Protocols—Companion Standard for Basic Telecontrol Tasks*

- *IEC 60870-5-102 Ed. 1.0: Companion Standard for the Transmission of Integrated Totals in Electric Power Systems*
- *IEC 60870-5-103 Ed. 1.0: Transmission Protocols—Companion Standard for the Informative Interface of Protection Equipment*
- *IEC 60870-5-104 Ed. 2.0: Transmission Protocols—Network Access for IEC 60870-5-101 Using Standard Transport Profiles*

The whole standard series is under continuous development and products are based on standards that are used in Europe, Asia, and the United States. They allow a vendor-independent communication among telecontrol and substation automation devices. Compared to the IEC 61850 standard series, the IEC 60870-5 does not offer the possibility to define typical devices in a standardized manner. So, in some cases IEC 608970-5 could be replaced by IEC 61850 standards, whereas only in a few situations does replacing one protocol with another protocol lead to additional values.

Beside the already mentioned basic and companion standards, the IEC 60870-5 series includes the following substandards dealing with testing:[18]

- *IEC 60870-5-6 Ed. 1.0: Guidelines for Conformance Testing for the IEC 60870-5 Companion Standards*
- *IEC/TS 60870-5-601 Ed. 1.0: Conformance Test Cases for the IEC 60870-5-101 Companion Standard*
- *IEC/TS 60870-5-604 Ed. 1.0: Conformance Test Cases for the IEC 60870-5-104 Companion Standard*

Because IEC 60870-5-101 and –104 are the most established companion standards, they are an explicit part of the SIA and described next. IEC 60870-5-102[43] is occasionally used, and IEC 60870-5-103[44] is used in various protection equipment.[23]

1.2.6.2.2 IEC 60870-5-101 and –104 IEC 60870-5-101[7] defines a communication profile that allows sending basic telecontrol messages between central telecontrol stations and telecontrol outstations. Permanent and directly connected data circuits between the stations are used. In some cases, several applications have to send the same type of messages between telecontrol stations. Therefore, data networks that contain relay stations could be used. These stations would store and forward the messages and provide only a virtual circuit instead of a physical one. Thus, the messages are variably delayed related to the network traffic load. The result is that it is not possible to use the link layer as it is defined in part 101. In special cases, however, it is possible to connect telecontrol stations that have all three layers specified in part 101 to suitable data networks by using stations of the packet assembler-disassembler (PAD) type, providing access for balanced communication. In all other cases, part 104[45] can be used to realize balanced access via a suitable transport protocol because it does not use the link functions of part 101. Hence, IEC 60870-5-104 includes a combination of the application layer defined in IEC 60870-5-101 and the transport functions from TCP/IP.

1.2.6.2.3 IEC 61850 Substation Devices   Working Groups 10, 17, and 18 in TC 57 are responsible for the IEC 61850 standard series, *Communication Networks and Systems in Substations*, which is one of the most used and recommended standard series for smart grids.[5] It aims at increasing interoperability between multivendor IEDs in substations, enabling data exchange and using data to implement the functionality required by the application. The IEEE (Institute of Electrical and Electronics Engineers) definition of *interoperability**[*] is used. So, it is not the goal to reach *interchangeability.*[†]

In addition to the communication technologies according to the single levels of the ISO/OSI layer, IEC 61850 comprises solutions for system aspects like project management; domain-specific data models including model extension methodologies; domain-specific services; a configuration language; and conformance tests. As in other standard series, the subparts of IEC 61850 have different focuses (e.g., IED configuration, device testing, data modeling, and abstract communication interfaces and their mapping on specific communication technologies).

---

[*] "Ability of a system or a product to work with other systems or products without special effort on the part of the customer. Interoperability is made possible by the implementation of standards" (http://www.ieee.org/education_careers/education/standards/standards_glossary.html).

[†] "Ability of a system or product to be compatible with or to be used in place of other systems or products without special effort by the user" (http://www.ieee.org/education_careers/education/standards/standards_glossary.html).

From a hierarchical perspective, a real physical device is modeled as a logical device (LD). Each LD consists of various logical nodes (LNs), described in IEC 61850-7-4.[46] Services conform to IEC 61850-7-2[47] and implementation of the abstract communication service interface (ACSI) is used for the communication with the LD. The IEDs themselves can be configured by Substation Configuration Language (SCL) files, described in IEC 61850-6.[48] Configuration issues could be networks, model entities, provided services, and integration into the grid.

The following standards are currently part of the standard series IEC 61850:[18]

- *IEC/TR 61850-1 Ed. 1.0: Introduction and Overview*
- *IEC/TS 61850-2 Ed. 1.0: Glossary*
- *IEC 61850-3 Ed. 1.0: General Requirements*
- *IEC 61850-4 Ed. 1.0: System and Project Management*
- *IEC 61850-5 Ed. 1.0: Communication Requirements for Functions and Device Models*
- *IEC 61850-6 Ed. 2.0: Configuration Description Language for Communication in Electrical Substations Related to IEDs*
- *IEC 61850-7-1 Ed. 1.0: Basic Communication Structure for Substation and Feeder Equipment—Principles and Models*
- *IEC 61850-7-2 Ed. 2.0: Basic Information and Communication Structure—Abstract Communication Service Interface (ACSI)*
- *IEC 61850-7-3 Ed. 2.0: Basic Communication Structure— Common Data Classes (CDCs)*
- *IEC 61850-7-4 Ed. 2.0: Basic Communication Structure— Compatible Logical Node Classes and Data Object Classes*
- *IEC 61850-7-410 Ed. 1.0: Hydroelectric Power Plants— Communication for Monitoring and Control*
- *IEC 61850-7-420 Ed. 1.0: Basic Communication Structure— Distributed Energy Resources Logical Nodes*
- *IEC 61850-8-1 Ed. 1.0: Specific Communication Service Mapping (SCSM)—Mappings to MMS (ISO 9506-1 and ISO 9506-2) and to ISO/IEC 8802-3*
- *IEC 61850-9-1 Ed. 1.0: Specific Communication Service Mapping (SCSM)—Sampled Values over Serial Unidirectional Multidrop Point to Point Link*

- *IEC 61850-9-2 Ed. 1.0: Specific Communication Service Mapping (SCSM)—Sampled Values over ISO/IEC 8802-3*
- *IEC 61850-10 Ed. 1.0: Conformance Testing*
- *IEC/TS 61850-80-1 Ed. 1.0: Guideline to Exchanging Information from a CDC-Based Data Model Using IEC 60870-5-101 or IEC 60870-5-104*
- *IEC/TR 61850-90-1 Ed. 1.0: Use of IEC 61850 for the Communication between Substations*

1.2.6.2.4 IEC 61850 Devices beyond the Substation   Whereas IEC 61850 was primarily intended to cope with substation automation, other devices were a later focus. Substandards IEC 61850-7-410[49] and –7-420[50] deal with those devices.

IEC 61850-7-410 includes extensions of the information model for hydroelectric power plants. The models define many LNs, which describe automation logic and thus go far beyond the IEC 61850-7-4 definitions. The main objective is to enable automation and monitoring of hydroelectric power plants in a way that could last for the next centuries. This is possible because during the next 10 to 20 years, hydro plant control and monitoring system will be renewed. Sustainable interoperability is of special interest in this area.[23]

IEC 6180-7-420 represents extensions for DER like photovoltaic, combined heat and power (CHP), fuel cells and reciprocating engines. In contrast to the field of substation automation, in which only a few global players control the market, many small and medium enterprises participate in the DER market. Hence, it is an important challenge to specify internationally accepted information models. In the future, these models will be tested for their practical suitability step by step.

Extensions of the IEC 61850 information model for wind power plants are part of the IEC 61400-25 standard.[9]

1.2.6.2.5 Communication Industry Standard Protocol Stacks   To realize a communication to the field devices, IT transport protocols must be used. During the last decades, some transport protocols, like TCP/IP and Ethernet, have been established. Hence, they were recommended for use in the utility domain.[2]

- **TCP:** The TCP is mainly based on two standards, RFC (Request for Comments) 793 (http://tools.ietf.org/html/rfc793) and RFC 1323 (http://tools.ietf.org/html/rfc1323). TCP specifies how data can be exchanged between two computers. It is supported and used by all recent operating systems. Furthermore, it is one of the core protocols of the Internet Protocol Suite, so that all major Internet applications like the World Wide Web and e-mail rely on it. TCP provides reliable, connection-oriented, and packet-switching communication.

- **IP:** RFC 791 (http://tools.ietf.org/html/rfc791) and RFC 2460 (http://tools.ietf.org/html/rfc2460) standardize the IP, a well-established network protocol within computer networks. It is also one of the core protocols of the Internet and allows the use of TCP. The main objective of IP is to route data packets across network boundaries, whereas the transmission from the source host to the destination host is solely based on their addresses.

- **Ethernet:** In the TCP/IP stack, Ethernet is the lowest layer, the basis for the IP. Ethernet specifies software and hardware for wired data networks, so that data exchange among devices within a LAN is possible. It contains various definitions for a number of wiring and signaling standards. Those standards cope with both the physical layer of the OSI networking model and the data link layer (common addressing format and a variety of Media Access Control procedures).

1.2.6.2.6 IEC 61850-8-1 Mapping to MMS   Specific Communication Service Mappings (SCSMs) are part of IEC 61850-8-1.[51] In this, substandard mappings of the abstract model to MMS (ISO/IEC 9506-1 and –2) and Ethernet (ISO/IEC 8802-3) are specified for communications within the whole substation. The information exchange, based on GOOSE (Generic Object Oriented Substation Event) and GSSE (Generic Substation Status Event) messages for real-time requirements like trigger signals and a client-server communication for SCADA functions, is also defined.

1.2.6.2.7 IEC 61850-7-2 ACSI   IEC 61850-7-2[47] defines a basic communication infrastructure for substation and feeder equipment focusing on ACSIs, including their descriptions. An ACSI is intended for use for applications in the utility domain that require real-time cooperation of IEDs. Furthermore, the ACSI is technology independent in terms of the underlying communication systems. The definitions of the ACSI include a hierarchical class model of the information that could be accessed by communication systems, services operating on these classes, and parameters linked to each service. The following communication services between clients and remote servers are in the scope of the substandard:

- real-time data access and retrieval
- device control
- event reporting and logging
- setting group control
- self-description of devices
- data typing and discovery of data types
- file transfer

1.2.6.2.8 IEC 61850-7-3 and –7-4 Object Models Devices like DER and substations modeled through IEC 61850 concepts are used by specific applications. Constructed, attributed classes and Common Data Classes (CDCs) are related to those applications and defined in IEC 61850-7-3.[52] IEC 61850-7-4[46] uses these CDCs to define compatible data object classes. The abstract definitions from part 7-2[47] are mapped to concrete object definitions, which are used for specific protocols like MMS. In detail, the following specifications are included in part 7-3:

- CDC for status information
- CDC for measured information
- CDC for control
- CDC status settings
- CDC analogue settings
- attribute types used in these CDCs

One of the pursued objectives of the standard series is to reach a high degree of interoperability. Therefore, all data objects—which could be mandatory, optional, or conditional—within the whole data model are strongly defined in terms of syntax and semantics. The semantic interoperability is achieved through names assigned to common LNs, their data objects are defined in part 7-4[46], and they are part of the class model specified in 7-1[53] and defined in 7-2.[47] The names are used to build a hierarchical object reference applied for communicating with IEDs in automation systems and in substations as well as on distribution feeders. Also, normative naming rules are defined to avoid private and thus maybe incompatible extensions of LNs and data object names. In addition, dedicated LNs are defined in other parts like IEC 61850-7-420[50] to model more specific devices like DER. Some LN features like data sets and logs are not modeled in part 7-4 but in part 7-2.

In addition to the descriptions of device models and functions of substations and feeder equipment, device models and functions for the following issues can be described:

- substation-to-substation information exchange
- substation-to-control-center information exchange
- power-plant-to-control-center information exchange
- information exchange for distributed generation
- information exchange for metering

1.2.6.2.9 Mapping to Web Services   Web services (http://www.w3.org/2002/ws/) specified by the World Wide Web Consortium (W3C) are software applications that communicate with each other using XML interfaces to send messages via Internet protocols. Each Web service is identifiable by its Uniform Resource Identifier (URI). There are three types of roles in a typical Web service system:

- service broker
- service provider
- service requester

The service provider uses the WSDL (Web Services Description Language) standard to provide its services to the service broker. In some cases, a small and local server is used to offer a service to register Web services via the UDDI (Universal Description, Discovery, and Integration) standard. The service requester can also use WSDL to communicate with the service broker. It queries the broker's repository to find a QoS (quality of service) or requirement-fitting service. In case of success, the service requester exchanges the data with the chosen service provider using the Simple Object Access Protocol (SOAP) standard, for example.

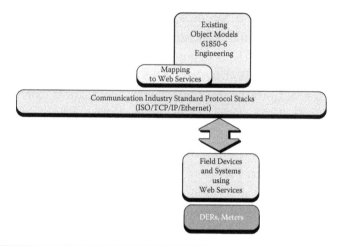

Figure 1.9   Pillar for DER and meters.

One example for mapping of an abstract communication to Web services is the IEC 61850 model, which is extended by IEC 61400-25-2[54] to enable modeling of wind power plants. In this context, another substandard was developed, IEC 61400-25-4[55] that specifies a Web-service-based communication for all IEC 61850-based data models.

1.2.6.3 DERs and Meters   Figure 1.9 illustrates the third pillar of the lower SIA part. This excerpt represents the Web-service-based communication for DER and some meter types. Therefore, an XML-based configuration language for substations called SCL are utilized just as protocols like TCP/IP are used. Two parts of this excerpt were discussed in Section 1.2.6.2.

1.2.6.3.1 Field Devices and Systems Using Web Services   The field devices shown in the other pillars are all accessible by standardized interfaces using standardized data models. This group, however, uses proprietary systems. For this reason, it is necessary to define interfaces so that it is possible to monitor and control the devices. Therefore, they define Web service interfaces to be connected to the upper layers.

1.2.6.3.2 Existing Object Models IEC 61850-6 Engineering   An important issue in standardization is the configuration of IEDs in substations; thus, IEC 61850-6[48] specifies a suitable description language, SCL, based on XML. By allowing the formal description of relations

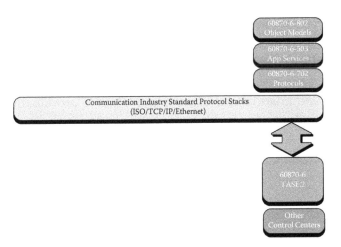

Figure 1.10   Pillar for other control center.

between automation systems and the processes like substations and switch yards, SCL is used to describe IED configurations and communication systems according to parts 5 and 7-x. From the application-level perspective, switch yard topologies and relations between their structure and SAS (Substation Automation System) functions configured on an IED can be described. The main objective of SCL is to enable an interoperable exchange of communication system configuration data between IED configuration tools and system configuration tools within a multivendor system architecture.

The definitions made in part 6 can be extended, or the use of values of objects can be restricted if it is necessary in terms of IEC 61850-8-1[51] and −9-2[56] concerning mappings of the abstract model defined in IEC 61950-7-x to specific communication technologies.

1.2.6.4 Other Control Centers   The last pillar (see Figure 1.10) deals with control centers that are not connected via IEC 61850, but via IEC 60870-6. Hence, a communication mainly based on TASE.2 is considered. The shown communication protocol stack was described in Section 1.2.6.2.

1.2.6.4.1 IEC 60870-6 TASE.2   TC 57 WG 7 is developing the IEC 60870-6 standard series *Telecontrol Protocols Compatible with ISO Standards and ITU-T Recommendations,* pursuing the goal of providing protocols that are able to run over WANs to interconnect control centers with heterogeneous databases and EMS applications. Those protocols and their services should be compliant to the OSI layered reference model and use existing ISO standards to the highest possible degree.

TASE.1 was the first published standard, and it was based on the ELCOM-90 protocol. The main objective was to provide the operation of an existing ELCOM-90 protocol over an OSI protocol stack. The TASE.1 API was developed as specified in the ELCOM-90 protocol documentation to enable replacements of the two protocols. The following substandards deal with TASE.1:[18]

- *IEC 60870-6-501 Ed. 1.0: TASE.1 Service Definitions*
- *IEC 60870-6-502 Ed. 1.0: TASE.1 Protocol Definitions*
- *IEC/TS 60870-6-504 Ed. 1.0: TASE.1 User Conventions*
- *IEC 60870-6-701 Ed. 1.0: Functional Profile for Providing the TASE.1 Application Service in End Systems*

TASE.2 was the successor of TASE.1 providing a utility-specific layer over MMS. It was developed for two major reasons: to provide extended functionalities and to maximize the use of existing OSI-compatible protocols like MMS. Whereas TASE.1 provides SCADA data and device control functionalities, TASE.2 also provides the exchange of information messages (e.g., short binary files) and structured data objects (e.g., transmission schedules). Therefore, a client-server architecture is used; its clients initiate transactions that are processed by the servers. Within the architecture, specific object models are used to define the transactions and services. In addition, the exchanged data were separately defined as static data objects. Hence, a distinction between the exchanged data and the used services was made. In addition to the object model, an anonymous point-oriented model is used to identify the received values and controlled devices. As for TASE.1, the following substandards deal with TASE.2:[18]

- *IEC 60870-6-503 Ed. 2.0: TASE.2 Services and Protocol*
- *IEC/TR 60870-6-505 Ed. 1.1 Consol. with am1: TASE.2 User Guide*
- *IEC 60870-6-702 Ed. 1.0: Functional Profile for Providing the TASE.2 Application Service in End Systems*
- *IEC 60870-6-802 Ed. 2.1 Consol. with am1: TASE.2 Object Models*

Beside the TASE.1 and TASE.2 specific sub-standards, the series comprises the following, more general parts:[18]

- *IEC/TR 60870-6-1 Ed. 1.0: Application Context and Organization of Standards*
- *IEC 60870-6-2 Ed. 1.0: Use of Basic Standards (OSI Layers 1–4)*
- *IEC 60870-6-601 Ed. 1.0: Functional Profile for Providing the Connection-Oriented Transport Service in an End System Connected via Permanent Access to a Packet Switched Data Network*
- *IEC/TS 60870-6-602 Ed. 1.0: TASE Transport Profiles*

1.2.6.4.2 IEC 60870-6-702 Protocols   IEC 60870-6-702[57] defines a functional profile covering the provision of the TASE.2 communication services between two control center end systems. Furthermore, the provision of the OSI connection mode presentation and session services between the end systems is defined by the functional profile.

1.2.6.4.3 IEC 60870-6-503 App Services   Part 6-503[58] of IEC 60870 defines the TASE.2 application modeling and service definitions. It specifies a method of exchanging time-critical control center data through WANs and LANs using fully ISO-compliant protocol stacks. Furthermore, it contains provisions for supporting both centralized and distributed architectures. It includes the exchange of real-time data indications, control operations, time series data, scheduling and accounting information, remote program control, and event notification. The use of TASE.2 is not restricted to control center data exchange. It may be applied in any other domain having comparable requirements. Examples of such domains are power plants, factory automation, and process control automation.

This standard does not specify individual implementations or products and does not constrain the implementation of entities and interfaces within a computer system. This standard specifies the externally visible functionality of implementations together with conformance requirements for such functionalities.

1.2.6.4.4 IEC 60870-6-802 Object Models   The primary objective of TASE.2 is transferring data between control systems and initiating control actions. Thereby, data is represented by object instances.

IEC 60870-6-802[59] proposes object models, representing objects for transfer, from which to define object instances. Local systems may not maintain a copy of every attribute of an object instance.

1.2.7 Security and Data Management

The SIA includes the IEC 62351 security standard as a cross section for data and communication security (in TC 57). It is drafted on the left side of Figure 1.1. The IEC 62351 includes eight parts: Part –1 provides a general introduction, and part –2 includes some definitions used in the standard. Parts –3 to –6 provide security enhancements for[15]

- profiles including TCP/IP (IEC 62351-3),
- profiles including MMS (IEC 62351-4),
- IEC 60870-5 and derivatives (IEC 62351-5), and
- IEC 61850 profiles (IEC 62351-6)

Part –7 of the standard is separately outlined in the SIA overview (see Figure 1.1) and deals with domain-specific data models for network management. An eighth part, which will consider role-based access control, is actually planned and not yet integrated in the SIA overview.

IEC 62351 is a standard for data and communication security. It is not a standard for information security management. Such security management methods can be found in IEC 62443 or, of course, the ISO/IEC 27k.

Next, we first explain the security enhancements defined in parts –3 to –6 and their benefits and restrictions. After that, we focus on the network management defined in IEC 62351-7. The last part of this section gives an overview of IEC 62351.

IEC 62351 parts –3 to –6 provide security enhancements described next.

1.2.7.1 Secure Communication via IEC 62351-3   IEC 62351-3 deals with the securing of TCP/IP -based protocols. The entire part –3 standard is about securing the communication on the transport layer through TLS (Transport Layer Security).[60] In general, TLS, as a successor of SSL (Secure Sockets Layer), realizes a secure communication through a hybrid encryption. Such an encryption makes use of asymmetric and symmetric encryption. The asymmetric encryption is used to securely

exchange symmetric keys, and the symmetric keys are used to encrypt the transferred data. The symmetric encryption is used because of its better performance. The asymmetric encryption, which only initializes the communication process as described, makes use of certificates. Server and client certificates are possible. In general, the authentication through a server-based certificate is very common. A certificate is a statement from a trusted third party (TTP) that includes a public key. The TTP guarantees that the included public key belongs to the certificate holder.

To conform to this standard, some aspects or parameters for the use of TLS must be mentioned.

- Only TLS version 1.0 (or at least SSL version 3.1) is allowed.
- MACs (message authentication codes) that are optional in TLS shall be used.
- Symmetric keys must be time-based negotiated by the calling nodes. For this cipher renegotiation call, there must be a time-out.
- For certification management, it is necessary to have more than one certification authority.
- The size of a certificate shall not be longer than 8,192 bytes.
- Certificate exchange shall be bidirectional.
- Certificate revocation is specified in RFC 3280.
- Signing via Rivest, Shamir, and Adleman (RSA) or digital signature standard (DSS) shall be supported.
- Key exchange with a maximum key size of 1,024 bits via RSA or Diffie-Hellman shall be supported.

The secured communication shall be on a separated port so that nonsecured communication can coexist. The use of this security enhancement provides some benefits for integrity, confidentiality, and authenticity. The protection goal authenticity is reached through the use of certificates. The encryption of the connections via TLS leads to confidentiality, and the use of MACs brings integrity. There are some restrictions for the use of TLS as a security enhancement. TLS does not mention the protection goal availability, so this standard will not protect against denial-of-service attacks.

1.2.7.2 Secure Profiles through IEC 62351-4   IEC 62351-4 brings mandatory and optional security enhancements for a secure communication

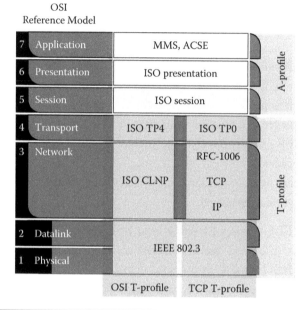

Figure 1.11 Profile security. (Reprinted with permission from International Electrotechnical Commission (IEC). International Electrotechnical Commission (IEC). *62351-7: Data and Communication Security Security through Network and System Management* (2007). Geneva, Switzerland: IEC.)

when using MMS (ISO/IEC 9506). IEC 61850-8-1 and IEC 60870-6 use MMS, either the OSI or TCP profiles, in a seven-layer connection-oriented mechanism, which is drafted in Figure 1.11. Therefore, different security profiles are considered as A and T profiles in this standard. Both can be found in the TC 57 context. The security profiles define protocols and requirements for the layers in the OSI reference model. The A profiles or application profiles are concerned with OSI layers five to seven, and the T profiles or transport profiles are pertinent to layers one to four. One can see these determinations on the right side of Figure 1.11. At the bottom of Figure 1.11, one can see a further distinction at the T profile into the OSI T profile and TCP T profile. The security of OSI T profiles is out of the scope of IEC 62351-4.[61]

An implementation of MMS must mention secure profiles to be compliant with this standard. There must be a possibility to use certificates for authentication. Furthermore, there must be a possibility to decide whether a secure or nonsecure profile is necessary for acceptance or initiation of communication or if it is not necessary. A secure security log is recommended. For peer authentication, association control service element (ACSE) (ISO 8650) shall use the ACSE authentication mechanism and authentication value fields. To be backward compatible, authentication values can be excluded for nonsecure profiles. A certificate-based MMS authentication value includes a signature certificate, a timestamp, and a signed value. Certificates must have a maximum size of 8,192 octets and shall be based on X.509. The signed value is a timestamp reduced with secure hash algorithm (SHA-1) and signed with RSA. If the sent timestamp differs from the encoded timestamp, the connection shall be refused. There are some other conditions for a connection abort. Messages older than 10 minutes will be ignored. So, there is a window of vulnerability of 10 minutes, during which the same signed value could be used by an attacker.

To be compliant to this standard, secure TCP T profiles must be used. In Figure 1.11, one can see the TCP T profile drafted on the right. From layers 3 to 1, the following protocols are mentioned in the TCP T profile: RFC 1006 (ISO Transport Service), TCP, IP, and IEEE 802.3 (Ethernet). This standard does not specify security specifications for these protocols or describe the use of TLS. It focuses on the layer 4 ISO TP0 protocol and specifies a secure RFC-1006 profile. This standard defines ports for the use of secure and nonsecure T profiles. The TLS defined in IEC 62351-3 shall be used. Furthermore, this standard defines things like transport protocol data unit (TPDU) to be ignored, size of transport selectors (TSELs), size of certificates, time to check certificate revocation, and recommended TLS cipher suites.

1.2.7.3 Authentication Technique of IEC 62351-5   Part –5 of the IECD 62351 standard deals with securing IEC 60870-5 protocols and derivates. It focuses on authentication mechanisms on the application layer. Security goals like confidentiality of data are out of the scope of this standard, but when IEC 60870-5-104 is in use, part –3 of this standard shall be mentioned.[62]

The protocols that shall be secured through this specification come with specific circumstances. The considered protocols of this security enhancement have an asymmetric communication and message orientation in common. There is a controlling and a controlled station, so we have a bidirectional communication. There are some security challenges with this kind of protocol, which avoids the adoption of some security mechanisms. Some of these challenges are missing or poor sequence numbers and missing or poor integrity mechanisms, limited frame length, long upgrade intervals, and more.

The authentication mechanism described in this specification makes use of a generic challenge-response concept, which shall be mapped into different standards. The key element for the authentication is a keyed hash message authentication code (HMAC). A HMAC is a MAC with a specific hash algorithm. By creating an HMAC, a hash value of a message is generated and then encrypted with a shared secret and symmetric key. The listener who also knows the secret can perform the hashing and encryption, so that the listener knows whether the message was modified and that only the other person with the secret key could have sent the message. An HMAC is not a digital signature because the MAC key is known by more than one person. Through hashing, manipulation of messages can be detected, so this accounts for integrity requirements. The use of a shared secret key for the HMAC between both sides gains authenticity. To reach a secure key exchange between the nodes, there are three different keys: an update key, a monitoring session key, and a control session key. The update key is used to encrypt the session keys. For security reasons, there are two different session keys for the monitoring and control direction. The update key is a preshared secret. The process to securely update unique keys (UKs) or a public key mechanism for that is out of the scope of this standard. The existence of an update key is a precondition for every node.

The authentication process specified in this standard is used for critical messages, but also for periodic messages. Before performing critical protocol messages, the application service data unit (ASDU), the executor of such a message, will initiate an authentication. The challenger of the critical ASDU has to start an authentication challenge, to which the executor will respond via an authentication response. Before authentication via an HMAC, it will be checked whether a common secret session key exists. The session key has to be transferred encrypted with the update key.

To be backward compatible, a nonsecure communication shall also be mentioned and possible. There are some further requirements like interoperability requirements and conformance statements for special applications within this standard.

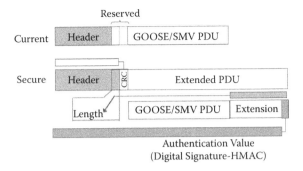

Figure 1.12 Extended PDU. (Reprinted with permission from International Electrotechnical Commission (IEC). *62351: Data and Communication Security* (2006). Geneva, Switzerland: IEC.)

1.2.7.4 PDU Security Extension of IEC 62351-6   Part –6 of IEC 62351 deals with the security of the IEC 61850 protocols. Profiles using MMS shall mention IEC 62351-4. Profiles using Simple Network Time Protocol (SNTP) should use RFC 2030 with authentication algorithms. This standard specifies PDU enhancements. The PDU shall include a MAC, which can be used for authentication. Figure 1.12 illustrates this extension.

An Application Protocol Data Unit (APDU) shall only be performed if the calculated MAC is identical to the sent MAC. Messages older than 2 minutes shall be ignored to avoid replay of GOOSE or sample measured value (SMV) messages.

To conform to this standard, SCL has to be enhanced. It must include certificates to realize authentication and encryption. The access point definition of SCL must include GOOSE-Security (IEC 61850-8-1) and SMV-Security (IEC 61850-9-2).

Encryption is not recommended for applications using GOOSE and IEC 61850-9-2 in combination with multicast because of the response time requirements.[63]

1.2.7.5 Intrusion Detection with IEC 62351-7   Part –7 of IEC 62351 is about network and system management for power systems. Therefore, it specifies abstract data models for controlling and monitoring the network and connected devices. The information of these data models shall be used as additional information for intrusion detection systems. The intention of this standard is to take availability requirements into account. The monitoring of the network and connected

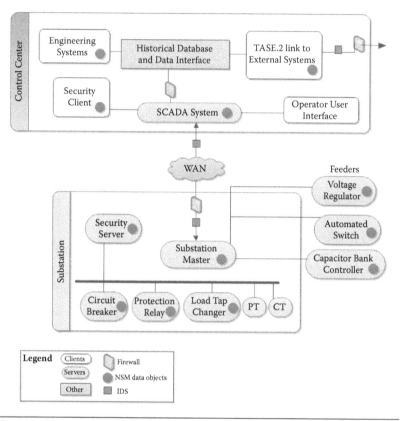

Figure 1.13   Security monitoring architecture of IEC 62351-7. CT, current transformer; IDS, intrusion detector system; NSM, network and system management; PT, potential transformer. (Reprinted with permission from International Electrotechnical Commission (IEC). International Electrotechnical Commission (IEC). *62351-7: Data and Communication Security Security through Network and System Management* (2007). Geneva, Switzerland: IEC.)

devices shall detect attacks. Also, the controlling of the network and connected devices shall react on an detected attack.[61] IEC 62351-7 does not define actions for alarms corresponding to these monitoring data models or specify the protocols to which the abstract data models could be mapped. In Figure 1.13, one can see the basic elements of a power system operation system and corresponding elements of the security-monitoring architecture of IEC 62351.[61]

1.3 Application of the SIA

At first, it should be clear that the SIA is not a step-by-step guide to build an ICT infrastructure in the energy domain, but a blueprint

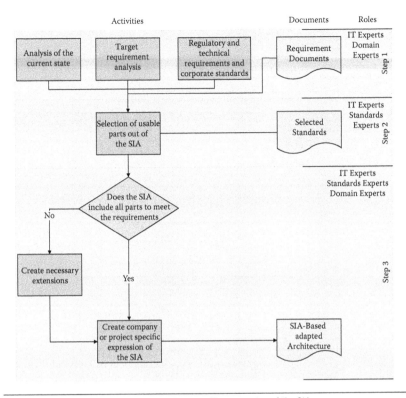

Figure 1.14 Create a company- or project-specific expression of the SIA.

that focuses on IEC-specific standards. To use the SIA, it is necessary to integrate the architecture in the company workflow or build up an entirely new process. The next section presents an example of how to use the SIA.

Figure 1.14 shows a rough procedure model on how to create a specific SIA that fits custom needs. The described procedure model is based on the specification of methods as described by Gutzwiller.[64] According to Gutzwiller, a method is described using the following elements: activity, role, specification document, meta model, and technique. Here, we present only a rough procedure model and therefore focus on activities (steps), roles, and specification documents. The steps are illustrated as flow diagrams enhanced by roles and specification documents.

The first goal should be the creation of a SIA-based architecture with specific adaptations. To achieve this, a thorough analysis of the current state as well as the target requirements is necessary (see step 1). In addition, the regulatory and technical requirements must be taken

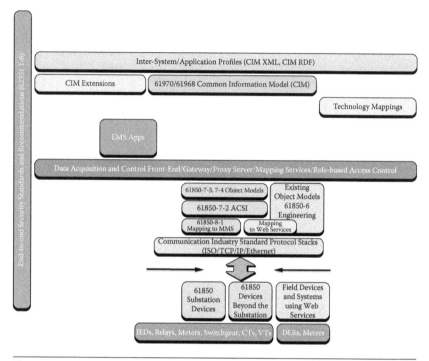

Figure 1.15 Fictional example of the selected SIA parts.

into account. Also, many companies have internal corporate standards that should contribute to the requirements.

With these previously identified specifications, the usable parts have to be selected from the original SIA (see Figure 1.15 for a fictional example), step 2 in Figure 1.14.

Step 3 includes the decision about the completeness of the cutout of the SIA. If one or more requirements are not met, it is necessary to extend the SIA with custom parts. These extensions could consist of other non-IEC standards, corporate standards, or regulatory rules. An example result document of an adapted architecture can be seen in Figure 1.16. The unused parts have been removed, and the used ones have been slightly shrunk together. At the bottom left, there is an example extension with the OPC UA IEC 62541 standard.

This adapted architecture can now be used as a starting point in new or migration projects. It is possible to construct requirements and road maps or even checklists to track the progress of adapting the SIA in the implementation process based on this.

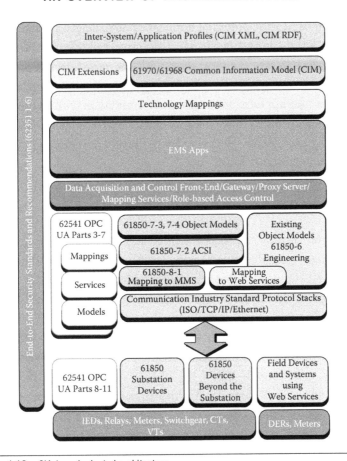

Figure 1.16 SIA-based adapted architecture.

One additional important point has to be taken into account. The SIA does not dictate the implementation of the interfaces between the standards across the layers. It is necessary to consult the specific standard documentation for implementation details for the recommended interfaces.

Furthermore, there is a lack of harmonization work between particular standards. This leads to an individual mapping with a certain degree of freedom, which results in unique characteristics.

The IEC is developing a smart grid mapping tool to support the creation process for an adapted architecture primarily for the identification and selection of parts from the SIA. It consists mainly of a metadata database that also contains information on the included standards and the direct or indirect connections to other standards within the SIA. One potential application is to select a starting standard, check the connections in all directions, and select the requested standards to reiterate the procedure with the new selected one to get the favored set of standards. Furthermore, the data include information about the application domains of standards within the energy domain like advanced metering infrastructure (AMI) or DER etc., so it should be possible to select the desired target domain and obtain all the relevant standards for it.

1.4 Summary and Outlook

This chapter showed the recommended standards necessary for a successful integration of technical smart grid infrastructures. The agreed-on and used standards solve the challenge of technical integration. In particular, the TC 57 Reference Architecture (IEC TR 62357) was examined in more detail. It provides an architecture and overview for standards of IEC TC 57, whose standards have also been referenced by various national and international smart grid standardization road maps (e.g., references 1–4).

In the TC 57 Reference Architecture, the integration of business partners was considered to connect the smart grid's associated businesses, such as the energy markets and utilities. Furthermore, standards for integration of field devices with SCADA systems were described, and details on which standards can be applied for the important cross-cutting concerns security and data management were provided.

The IEC TR 62357 Reference Architecture provides a comprehensive framework of standards to integrate various smart grid participants. Currently, standards need to be harmonized to achieve a seamless integration, which is already one of the TC's key objectives. A special issue being addressed at present is the harmonization between IEC 61970/61968 (CIM) and IEC 61850. WG 19, originally founded to resolve model differences where there is an overlap between standards and to develop a vision for TC 57 for the future architecture, has already treated this issue to some degree. Despite harmonization efforts, modifications to standards themselves might become necessary, as for instance proposed for IEC CIM 61970/61968,[20] or new standards might emerge, which has to be considered in dependent standards. In addition, other standards that are out of the TC's scope, are to be considered for integration, like the OPC UA, shown in the example in Section 1.3. According to this approach, the reference architecture has to evolve constantly and incorporate these changes.

The future vision for the reference architecture establishes a basis for seamless integration of technical systems involved in the smart grid and thus moves closer to realization of the vision of the future power grid. Recognizing that a single, agreed information model (CIM) can avoid mappings and inconsistencies between standards and, beyond that, is open to and linked with other related TCs and industry consortiums will clearly influence the current picture of the SIA.

References

1. Deutsche Kommission Elektrotechnik (DKE). *Die deutsche Normungsroadmap E-Energy/Smart Grid* (2010). Frankfurt, Germany: VDE.
2. National Institute for Standards and Technology. *NIST Framework and Roadmap for Smart Grid Interoperability Standards*, Release 1.0 (2010). http://www.nist.gov/public_affairs/releases/upload/smartgrid_interope rability_final.pdf
3. *Japan's Roadmap to International Standardization for Smart Grid and Collaborations with Other Countries* (2010).
4. State Grid China. *SGCC Framework and Roadmap for Strong and Smart Grid Standards* (2010).
5. S. Rohjans, M. Uslar, R. Bleiker, J. González, M. Specht, T. Suding, and T. Weidelt. Survey of Smart Grid Standardization Studies and Recommendations. In *First IEEE International Conference on Smart Grid Communications* (2010), pp. 583–588.
6. M. Uslar, S. Rohjans, R. Bleiker, J. M. González, T. Suding, M. Specht, and T. Weidelt. Survey of Smart Grid Standardization Studies and Recommendations—Part 2. In *First IEEE International Conference on Smart Grid Communications* (2010), pp. 1–6.
7. International Electrotechnical Commission (IEC). *60870-5-101 Ed. 2.0: Telecontrol Equipment and Systems—Part 5-101: Transmission Protocols—Companion Standard for Basic Telecontrol Tasks* (2003). Geneva, Switzerland: IEC.
8. International Electrotechnical Commission (IEC). *61334-1-1 Ed. 1.0: Distribution Automation Using Distribution Line Carrier Systems—Part 1: General Considerations—Section 1: Distribution Automation System Architecture* (1995). Geneva, Switzerland: IEC.

9. International Electrotechnical Commission (IEC). *61400-25-1 Ed. 1.0: Wind Turbines—Part 25-1: Communications for Monitoring and Control of Wind Power Plants—Overall Description of Principles and Models* (2006). Geneva, Switzerland: IEC.

10. International Electrotechnical Commission (IEC). *61850-1 Ed. 1.0: Communication Networks and Systems in Substations—Part 1: Introduction and Overview* (2003). Geneva, Switzerland: IEC.

11. International Electrotechnical Commission (IEC). *61968-1: Application Integration at Electric Utilities—System Interfaces for Distribution Management Part 1: Interface Architecture and General Requirements* (2007). Geneva, Switzerland: IEC.

12. International Electrotechnical Commission (IEC). *61970-1 Ed. 1: Energy Management System Application Program Interface (EMS-API)— Part 1: Guidelines and General Requirements* (January 2005). Geneva, Switzerland: IEC.

13. International Electrotechnical Commission (IEC). *62056-21 Ed. 1.0: Electricity Metering—Data Exchange for Meter Reading, Tariff and Load Control—Part 21: Direct Local Data Exchange* (2002). Geneva, Switzerland: IEC.

14. International Electrotechnical Commission (IEC). *62325-101 DTR Ed. 1: Framework for Energy Market Communications Part 101: General Guidelines and Requirements* (January 2004). Geneva, Switzerland: IEC.

15. International Electrotechnical Commission (IEC). *62351: Data and Communication Security* (2006). Geneva, Switzerland: IEC.

16. International Electrotechnical Commission (IEC). *IEC 62357 Second Edition: TC 57 Architecture—Part 1: Reference Architecture for TC 57— Draft* (2009). Geneva, Switzerland: IEC.

17. International Electrotechnical Commission (IEC). *61970-2 Ed. 1: Energy Management System Application Program Interface (EMS-API)— Part 2: Glossary* (January 2003). Geneva, Switzerland: IEC.

18. International Electrotechnical Commission (IEC). Webstore International Electrotechnical Commission (2011). http://webstore.iec.ch (accessed March 30, 2011).

19. Electric Power Research Institute (EPRI). *An Introduction to the CIM for Integrating Distribution* (2008). Palo Alto, CA: EPRI.

20. M. Uslar, S. Rohjans, M. Specht, and J. Gonzales. What Is the CIM lacking? In *First IEEE International Conference on Smart Grid Communications* (2010), pp. 1–8.

21. International Electrotechnical Commission (IEC). *61970-452: Energy Management System Application Program Interface (EMS-API)—Part 452: CIM Transmission Network Model Exchange Profile* (2009). Geneva, Switzerland: IEC.

22. International Electrotechnical Commission (IEC). *61968-13 Ed. 1: Application Integration at Electric Utilities—System Interfaces for Distribution Management—Part 13: CIM RDF Model Exchange Format for Distribution* (2008). Geneva, Switzerland: IEC.

23. OFFIS, SCC Consulting, and M. Management Coaching. *Untersuchung des Normungsumfeldes zum BMWi-Förderschwerpunkt 'E-Energy—IKT-basiertes Energiesystem der Zukunft'* (2009).

24. Open Management Group (OMG). *MDA Guide Version 1.0.1* (2003).

25. International Electrotechnical Commission (IEC). *61968-1: Application Integration at Electric Utilities—System Interfaces for Distribution Management Part 1: Interface Architecture and General Requirements (Draft)* (2010). Geneva, Switzerland: IEC.

26. International Electrotechnical Commission (IEC). *62056-46 Ed. 1.1 Consol. with am1: Electricity Metering—Data Exchange for Meter Reading, Tariff and Load Control—Part 46: Data Link Layer Using HDLC Protocol* (2007). Geneva, Switzerland: IEC.

27. International Electrotechnical Commission (IEC). *62056-53 Ed. 2.0: Electricity Metering—Data Exchange for Meter Reading, Tariff and Load Control—Part 53: COSEM Application Layer* (2006). Geneva, Switzerland: IEC.

28. International Electrotechnical Commission (IEC). *62056-42 Ed. 1.0: Electricity Metering—Data Exchange for Meter Reading, Tariff and Load Control—Part 42: Physical Layer Services and Procedures for Connection-Oriented Asynchronous Data Exchange* (2002). Geneva, Switzerland: IEC.

29. International Electrotechnical Commission (IEC). *62056-61 Ed. 2.0: Electricity Metering—Data Exchange for Meter Reading, Tariff and Load Control—Part 61: Object Identification System (OBIS)* (2006). Geneva, Switzerland: IEC.

30. International Electrotechnical Commission (IEC). *62056-62 Ed. 2.0: Electricity Metering—Data Exchange for Meter Reading, Tariff and Load Control—Part 62: Interface Classes* (2006). Geneva, Switzerland: IEC.

31. International Electrotechnical Commission (IEC). *62051 Ed. 1.0: Electricity Metering—Glossary of Terms* (1999). Geneva, Switzerland: IEC.

32. International Electrotechnical Commission (IEC). *61334-4-1 Ed. 1.0: Distribution Automation Using Distribution Line Carrier Systems—Part 4: Data Communication Protocols—Section 1: Reference Model of the Communication System* (1996). Geneva, Switzerland: IEC.

33. International Electrotechnical Commission (IEC). *61334-6 Ed. 1.0: Distribution Automation Using Distribution Line Carrier Systems—Part 6: A-XDR Encoding Rule* (2000). Geneva, Switzerland: IEC.

34. International Electrotechnical Commission (IEC). *61334-5-1 Ed. 2.0: Distribution Automation Using Distribution Line Carrier Systems—Part 5-1: Lower Layer Profiles—The Spread Frequency Shift Keying (S-FSK) Profile* (2001). Geneva, Switzerland: IEC.

35. International Electrotechnical Commission (IEC). *61334-5-2 Ed. 1.0: Distribution Automation Using Distribution Line Carrier Systems—Part 5-2: Lower Layer Profiles—Frequency Shift Keying (FSK) Profile* (1998). Geneva, Switzerland: IEC.

36. International Electrotechnical Commission (IEC). *61334-5-3 Ed. 1.0: Distribution Automation Using Distribution Line Carrier Systems—Part 5-3: Lower-Layer Profiles—Spread Spectrum Adaptive Wideband (SS-AW) Profile* (2001). Geneva, Switzerland: IEC.

37. International Electrotechnical Commission (IEC). *61334-5-4 Ed. 1.0: Distribution Automation Using Distribution Line Carrier Systems—Part 5-4: Lower Layer Profiles—Multi-carrier Modulation (MCM) Profile* (2001). Geneva, Switzerland: IEC.

38. International Electrotechnical Commission (IEC). *61334-5-5 Ed. 1.0: Distribution Automation Using Distribution Line Carrier Systems—Part 5-5: Lower Layer Profiles—Spread Spectrum—Fast Frequency Hopping (SS-FFH) Profile* (2001). Geneva, Switzerland: IEC.

39. International Electrotechnical Commission (IEC). *61334-4-511 Ed. 1.0: Distribution Automation Using Distribution Line Carrier Systems—Part 4-511: Data Communication Protocols—Systems Management—CIASE Protocol* (2000). Geneva, Switzerland: IEC.

40. International Electrotechnical Commission (IEC). *61334-4-512 Ed. 1.0: Distribution Automation Using Distribution Line Carrier Systems—Part 4-512: Data Communication Protocols—System Management Using Profile 61334-5-1—Management Information Base (MIB)* (2001). Geneva, Switzerland: IEC.

41. International Electrotechnical Commission (IEC). *61334-3-21 Ed. 1.0: Distribution Automation Using Distribution Line Carrier Systems—Part 3: Mains Signalling Requirements—Section 21: MV Phase-to-Phase Isolated Capacitive Coupling Device* (1996). Geneva, Switzerland: IEC.

42. International Electrotechnical Commission (IEC). *61334-3-22 Ed. 1.0: Distribution Automation Using Distribution Line Carrier Systems—Part 3-22: Mains Signalling Requirements—MV Phase-to-Earth and Screen-to-Earth Intrusive Coupling Devices* (2001). Geneva, Switzerland: IEC.

43. International Electrotechnical Commission (IEC). *60870-5-102 Ed. 1.0: Telecontrol Equipment and Systems—Part 5: Transmission Protocols—Section 102: Companion Standard for the Transmission of Integrated Totals in Electric Power Systems* (1996). Geneva, Switzerland: IEC.

44. International Electrotechnical Commission (IEC). *60870-5-103 Ed. 1.0: Telecontrol Equipment and Systems—Part 5-103: Transmission Protocols—Companion Standard for the Informative Interface of Protection Equipment* (1997). Geneva, Switzerland: IEC.

45. International Electrotechnical Commission (IEC). *60870-5-104 Ed. 2.0: Telecontrol Equipment and Systems—Part 5-104: Transmission Protocols—Network Access for IEC 60870-5-101 Using Standard Transport Profiles* (2006). Geneva, Switzerland: IEC.

46. International Electrotechnical Commission (IEC). *61850-7-4 Ed. 2.0: Communication Networks and Systems for Power Utility Automation—Part 7-4: Basic Communication Structure—Compatible Logical Node Classes and Data Object Classes* (2010). Geneva, Switzerland: IEC.

47. International Electrotechnical Commission (IEC). *61850-7-2 Ed. 2.0: Communication Networks and Systems for Power Utility Automation—Part 7-2: Basic Information and Communication Structure—Abstract Communication Service Interface (ACSI)* (2010). Geneva, Switzerland: IEC.

48. International Electrotechnical Commission (IEC). *61850-6 Ed. 2.0: Communication Networks and Systems for Power Utility Automation—Part 6: Configuration Description Language for Communication in Electrical Substations Related to IEDs* (2009). Geneva, Switzerland: IEC.

49. International Electrotechnical Commission (IEC). *61850-7-410 Ed. 1.0: Communication Networks and Systems for Power Utility Automation—Part 7-410: Hydroelectric Power Plants—Communication for Monitoring and Control* (2007). Geneva, Switzerland: IEC.

50. International Electrotechnical Commission (IEC). *61850-7-420 Ed. 1.0: Communication Networks and Systems for Power Utility Automation—Part 7-420: Basic Communication Structure—Distributed Energy Resources Logical Nodes* (2009). Geneva, Switzerland: IEC.

51. International Electrotechnical Commission (IEC). *61850-8-1 Ed. 1.0: Communication Networks and Systems in Substations—Part 8-1: Specific Communication Service Mapping (SCSM)—Mappings to MMS (ISO 9506-1 and ISO 9506-2) and to ISO/IEC 8802-3* (2004). Geneva, Switzerland: IEC.

52. International Electrotechnical Commission (IEC). *61850-7-3 Ed. 2.0: Communication Networks and Systems for Power Utility Automation—Part 7-3: Basic Communication Structure—Common Data Classes* (2010). Geneva, Switzerland: IEC.

53. International Electrotechnical Commission (IEC). *61850-7-1 Ed. 1.0: Communication Networks and Systems in Substations—Part 7-1: Basic Communication Structure for Substation and Feeder Equipment—Principles and Models* (2003). Geneva, Switzerland: IEC.

54. International Electrotechnical Commission (IEC). *61400-25-2 Ed. 1.0: Wind Turbines—Part 25-2: Communications for Monitoring and Control of Wind Power Plants—Information Models* (2006). Geneva, Switzerland: IEC.

55. International Electrotechnical Commission (IEC). *61400-25-4 Ed. 1.0: Wind Turbines—Part 25-4: Communications for Monitoring and Control of Wind Power Plants—Mapping to Communication Profile* (2008). Geneva, Switzerland: IEC.

56. International Electrotechnical Commission (IEC). *61850-9-2 Ed. 1.0: Communication Networks and Systems in Substations—Part 9-2: Specific Communication Service Mapping (SCSM)—Sampled Values over ISO/IEC 8802-3* (2004). Geneva, Switzerland: IEC.

57. International Electrotechnical Commission (IEC). *60870-6-702 Ed. 1.0: Telecontrol Equipment and Systems—Part 6-702: Telecontrol Protocols Compatible with ISO Standards and ITU-T Recommendations—Functional Profile for Providing the TASE.2 Application Service in End Systems* (1998). Geneva, Switzerland: IEC.

58. International Electrotechnical Commission (IEC). *60870-6-503 Ed. 2.0: Telecontrol Equipment and Systems—Part 6-503: Telecontrol Protocols Compatible with ISO Standards and ITU-T Recommendations—TASE.2 Services and Protocol* (2002). Geneva, Switzerland: IEC.

59. International Electrotechnical Commission (IEC). *60870-6-802 Ed. 2.1 Consol. with am1: Telecontrol equipment and Systems—Part 6-802: Telecontrol Protocols Compatible with ISO Standards and ITU-T Recommendations—TASE.2 Object Models* (2005). Geneva, Switzerland: IEC.

60. International Electrotechnical Commission (IEC). International Electrotechnical Commission (IEC). *62351-3: Data and Communication Security Profiles Including TCP/IP* (2005). Geneva, Switzerland: IEC.

61. International Electrotechnical Commission (IEC). International Electrotechnical Commission (IEC). *62351-7: Data and Communication Security Security through Network and System Management* (2007). Geneva, Switzerland: IEC.

62. International Electrotechnical Commission (IEC). *IEC 62351-5: Data and Communication Security Security for IEC 60870-5 and Derivatives* (2007). Geneva, Switzerland: IEC.

63. International Electrotechnical Commission (IEC). *IEC 62351-6: Data and Communication Security Security for IEC 61850 Profiles* (2005). Geneva, Switzerland: IEC.

64. T. Gutzwiller. Das CC RIM-Referenzmodell für den Entwurf von betrieblichen, transaktionsorientierten Informationssystemen. PhD thesis, Hochschule St. Gallen für Wirtschafts-, Rechts- und Sozialwissenschaften, St. Gallen, Switzerland (1994).

2

Smart Grid and Cloud Computing

Minimizing Power Consumption and Utility Expenditure in Data Centers

SUMIT KUMAR BOSE, MICHAEL SALSBURG, SCOTT BROCK, AND RONALD SKEOCH

Contents

Today's "Internet-scale" systems may be made up of several hundred or thousand servers spread across many geographies. These systems consume several megawatts of electricity a day. It is important therefore to build systems that are optimized for power management. However, building such a system is a challenge as trade-offs between application performance and power consumption need to be considered. In this chapter, we discuss recent advancements in cloud computing and smart grid technologies to design a power management system that helps reduce the power expenditure incurred by a cloud provider without "overtly" sacrificing the performance of the applications hosted by it. In particular, this chapter discusses ways in which a cloud provider can respond to various dynamic pricing signals received by the smart meters installed at its facilities, called data centers, by autonomously moving "noncritical" applications to remote sites during peak electric grid load situations by leveraging techniques from cloud computing.

2.1 Introduction

Today's "Internet-scale" systems are housed in geographically distributed server farms, typically known as data centers. These data centers may contain several hundred or thousand servers and are among the largest consumers of electricity. It is important in such scenarios to monitor not only the cost of managing the information technology (IT) infrastructure but also the cost of powering the IT infrastructure, also called the energy cost. It is estimated that the power expenditure is nearly one-fourth of the total operational cost of modern data centers. For example, the power consumption in data centers accounts for 1.2% of the overall electricity consumption in the United States and is projected to keep growing at 18% every year.[1] In light of these growing statistics, it is important to profile and infer the power utilization characteristics of applications and execute them in an efficient manner. Numerous research works in the past have explored strategies for efficient execution of applications with the aim of minimizing power consumption.[2–4]

In the following paragraphs, we discuss ways to monitor and manage the power consumption of applications during peak power grid load-occurring situations. Monitoring and managing power consumption at peak power grid load-occurring situations is crucial as the electricity expenditure during a peak power grid load-occurring situation could be overwhelmingly large compared to the total electricity expenditure during nonpeak load situations. The reason behind this is partly due to temporal variation in electricity prices: The electricity price in peak power grid situations is high due to demand-supply mismatch and the high cost of generating electricity at high loads. Information about dynamic pricing is communicated by the power utilities distribution companies to their consumers using smart meters and advanced metering infrastructures called smart grids as part of demand response (DR) programs.

With the help of smart meters installed at different data centers of a cloud provider, the power distribution companies can remotely monitor the electricity consumption at these centers. These power distribution companies can then make use of these advanced metering systems, if required, to push appropriate DR signals to data centers when faced with power shortages. From the perspective of a cloud provider, the DR signals received by them provide the necessary pricing information and indicate the prices that the power distribution company will charge the cloud provider for consuming electricity during periods of peak power grid load. In addition, it may specify a penalty that the cloud provider will incur if it fails to fulfill its commitment of curtailing its electricity consumption during these periods. To reduce its electricity consumption during such situations, a cloud provider needs to identify a subset of applications that it can afford to operate at suboptimal performance levels for brief durations and another subset of applications that it can afford to migrate to remote cloud locations. In the following paragraphs, we discuss the recent advancements in virtual machine (VM) migration technologies within cloud computing[5] and how these advancements can be leveraged to achieve this objective. This intelligent migration of VMs across different virtualized data centers in an autonomic manner helps to minimize power consumption during peak power grid load situations with minimal impact on application performance.

The chapter is organized as follows: Section 2.2 discusses the service-level agreements (SLAs) and the application assortment problem. Section 2.3 discusses the server virtualization and the cloud computing technology that enable seamless movement of applications from one data center to another. Section 2.4 outlines the detailed solution architecture and describes the interaction of the different solution components. Section 2.5 discusses the various components of the architecture at length and develops appropriate mathematical models for each of the components.

2.2 Service-Level Agreements

Before the service engagement between a cloud provider and a cloud consumer can begin, the two parties must mutually agree to the provisions of a legally enforceable service contract called the service-level agreement (SLA). This service contract is embodied in a document and formally defines the minimum performance criteria against which the service levels and hence the performance of a service provider will be compared. Further, the service contract lists the penalties for situations when the service provider fails to meet the obligations as committed by it prior to the initiation of the service engagement and when the performance falls below the promised standard. Broadly, an SLA can be of two types: infrastructure and application. Provisions within an infrastructure SLA are meant to indicate that the service parameters, such as the availability of the hardware and the networking switches, are the responsibility of the service provider. Provisions within an application SLA are meant to indicate that the service parameters, such as guaranteeing the response time and the throughput, are the responsibility of the service provider. Individual provisions of an SLA are known as service-level objectives (SLOs). The focus of this chapter is on application SLAs and their SLOs.

The service parameters, such as response time and throughput, are known as the performance metric. Typically, an application SLA specifies two quality indicators for any performance metric: *average* and *threshold*. If the performance metric under consideration is response time, then the threshold value indicates the maximum time that a service provider can take to service each individual user request. If the performance metric under consideration is throughput, then the threshold value indicates the minimum number of user requests that a service provider should be able to service within a given time window. Thus, the threshold value of a performance metric is the hard limit that, when breached, results in harmful consequences for both the cloud provider and the cloud consumer. The average value of a performance metric is an indicator of the desirable quality and level of service over relatively long periods of time. In subsequent discussions, we describe the analytic problem using response time as the key performance metric. The analysis can be extrapolated easily to other performance metrics.

Assume that the SLA for an application i requires the average response time to be R_{avg} and the threshold response time to be within R_{max}. We assume that the total electric load that the data center needs to shed, as communicated by the smart meter, is $d > 0$. Let P and P' ($P' < P$) indicate the power consumption of an application at the workload λ for achieving response times of R_{avg} and R_{max} ($R_{avg} < R_{max}$), respectively. Then, the objective is to exploit the difference in values of these two SLA parameters ($R_{max} - R_{avg}$) for every application so that the total electricity consumption by the applications at the data center during peak electric grid load can be curtailed by d. That is, the shedding in electricity consumption by the applications should not overtly affect the performance of the application. Thus, the quality of service as agreed to by the cloud service providers for the applications in their respective application SLAs should remain acceptable despite the reduction in electricity drawn by the applications. Earlier researchers have explored power performance trade-offs and dynamic voltage and frequency scaling (DVFS) schemes for maintaining the response time at a desired level with varying workloads. Thus, for workloads λ_1 and λ_2 such that $\lambda_1 < \lambda_2$ and a processor operating at a fixed frequency f, the request response times will be R_1 and R_2, respectively, with $R_1 < R_2$. These works therefore lower the operating frequency of the processor to f' ($f' < f$) such that $R_1 \approx R_1 \approx R_{avg}$. In contrast to these works, this chapter shows how to exploit the difference between R_{max} and R_{avg} to identify applications that are best suited for migration to remote sites so that the power expenditure due to high electricity cost at peak power grid situations can be reduced. In the following paragraphs, whenever an application is allocated barely enough computing resources such that the response time of the requests is R_{max}, the application is said to be operating at *threshold* SLA levels. However, when the application is allocated sufficient computing resources so that the response time of the requests is R_{avg}, the application is said to be operating at *standard* SLA levels.

A data center acts as a host to many different applications with varying input/output (I/O), central processing unit (CPU), and memory characteristics. In addition, each of these applications has a varying degree of tenacity to operate at threshold SLA levels during different times of the day. Thus, at time t_1 an application i_1 can operate at threshold SLA levels only for duration τ_{i_1}. At time t_2, the application can operate at threshold SLA levels for duration τ_{i_2}, $\tau_{i_2} \neq \tau_{i_1}$. Again, at time t_2 a different application i' can operate at threshold SLA levels for duration τ'_{i_2}. The problem then is to identify the following for a data center experiencing peak electricity load:

1. A candidate set of applications that can operate at threshold SLA levels for a fraction of duration for which the electric grid is experiencing peak load.
2. A candidate set of applications that need to be moved to another cloud site that is not experiencing an adverse electric grid load situation.

A combination of items 1 and 2 should ensure that the total curtailment in electricity drawn by the applications hosted at the data center should be at least d. We call this the application assortment problem. However, moving an application at run time from one physical machine on which it is already executing to another physical machine is fraught with challenges that need to be resolved.[6,7,8] Encapsulating applications within VMs, and moving entire VMs from one physical machine to another physical machine has been proposed as a way out of these challenges. The following section discusses live migration of VMs in detail.

2.3 Live Migration of a VM Image in Cloud Computing

Migration of VM images between geographically dispersed nodes in the distributed data center or the cloud requires, at a high level, that two major considerations be addressed. The first consideration is the duplication or replication of the guest data, which include storage and memory resident data. The second consideration is the transition or redirection of network communications from one logical network to another, ensuring that traffic will continue to reach the VM at its new location.

2.3.1 Data Migration

Replication of data on disk can be implemented in a number of ways, but most methods intercept writes from a VM guest to its "disk(s)." The most common approach is to encapsulate this service within a driver and to place the driver in the I/O stack of the host server that contains the guest. Another approach, and perhaps best from a performance perspective, is to use a replication solution that is located outside the server infrastructure that resides in components such as the storage array or storage area network (SAN) switch.

In the I/O driver approach, this driver is located in the I/O stack, usually in the device stack (just above the multipathing driver if it is present), on the host server. The driver is capable of reading the header information of each I/O frame to determine which writes to ignore and which to replicate. The driver is configurable so that any number of local drives can be replicated.

When the replication function resides in other infrastructure components, the I/O driver functionality effectively resides in the operating system (OS) of that component (SAN-OS, switch OS) so the distinction between the approaches is really the difference in location of the driver.

If a write is made to a disk that is to be replicated, that write is copied, the original is passed to its intended target, and the copy is routed to be delivered eventually to the remote storage where the remote VM host server resides.

One of the practical considerations in minimizing the time that it takes to migrate a VM between locations is the question of the amount of data that remains to be transferred over the network connection at the point in time the migration process is initiated. This of course depends on how different, in terms of changes that have to be applied, the local and remote VM images are. Replication can be implemented as an on-demand service or as a background service. Replication on demand has the advantage of not impacting network resources until a migration is needed. However, when the migration process is initiated, replication starts only at that point in time, so there is likely a relatively large quantity data that will need to be transferred, at least when compared to the alternative of background replication. With replication running as a background process, some network resources will be consumed during frequent regular intervals prior to receiving the migration command, but since the difference in local and remote images is likely to require fewer updates, the amount of data remaining to be transferred when the migration command occurs is significantly less. The background process will require a larger total amount of data to be transferred because incremental updates, as opposed to one single update, will have applied updates for some files multiple times. Thus, the choice is effectively a trade-off between a background process that prepositions as much data as possible to minimize the likely differences between local and remote images at the cost of some additional network loading or an on-demand process that most efficiently uses the network connection overall at the cost of a longer time interval to update the remote image.

However, regardless of the replication approach, there remains the task of replicating local resident data to the remote site. During a live migration of a VM guest, when the bulk of the disk data has been replicated and those data are nearly synchronous between the local and remote images, the local host begins to transfer data in local memory to the remote VM. At the end of this memory echo process, the source VM is paused, last-second disk updates are copied, and the final data resident in the local guest memory are copied to the remote site. It takes only a few tens of milliseconds to make the final transition from the source VM to the target VM instance. When the data transition is complete, restoration of network connectivity is required before the new VM image can resume its hosting function, completing the live migration process.

2.3.2 Network Migration

Once the remote VM image residing on the remote storage is completely duplicated by the replication process, network connectivity must be either redirected or reestablished to be accessible. At this point in the process, all of the network traffic is still being routed to the network where the "old" VM image is now paused. There are various means to restore network connectivity to the new VM image; two common approaches will suffice as examples of how this is accomplished.

One approach requires an Internet Protocol (IP) tunnel to be created from the old network address to the new guest. This IP tunnel is created just before the old VM is placed into its paused state. This tunnel allows network traffic inbound to the old address to reach the guest at its new location. As the new guest goes online, it will register its new address with the DNS (Domain Name System) server, and eventually when the DNS entries are updated any new client traffic will be able to connect with the guest directly. When connections over the IP tunnel are all closed, the tunnel will collapse. This mechanism allows migration of the guest from the old network to the new network without significant client disruption.

Another approach requires the creation of a Multi-Protocol Label Switching-Virtual Private Network (MPSL-VPN) mesh framework between the various sites. Once this framework is in place, all of the various sites exist as if they were all on a local network. Because the entire network environment is treated as a single entity, an Address Resolution Protocol (ARP) update to one switch makes the connections of the new VM available to the entire environment and accessible to any client. From a migration perspective, this means that as a guest comes online and receives a new IP address, this new address will be placed into the ARP tables, and immediately all traffic will be routed to this new location.

To accomplish the migration of VMs over geographic distances, it is necessary to have means to replicate a VM image and supporting volumes from the primary location to the secondary location and then to be able to restore network connectivity to the replicated image.

The choice of where to implement write splitting and the replication function depends primarily on whether the small additional processing and I/O load on the host is tolerable and does not significantly impact the host's performance. A host driver tends to be simpler to implement and less costly but does have a footprint on the host. SAN or switch-based implementations do not have an impact on the host but are typically more complex and costly.

Various methods are available to migrate network connectivity, but the methods used usually depend on selecting a method compatible with the existing infrastructure. The IP tunneling methods are simple but in some situations can cause delays for some clients until DNS caches are fully updated. The MPLS-VPN method is a more complex process but provides comparatively faster updates and improved performance.

2.4 Architecture

The solution architecture for the application assortment problem is shown in Figure 2.1. The architecture consists of three components: application manager, site broker, and hybrid cloud broker (HCB). The functionality of each of the three components is described next.

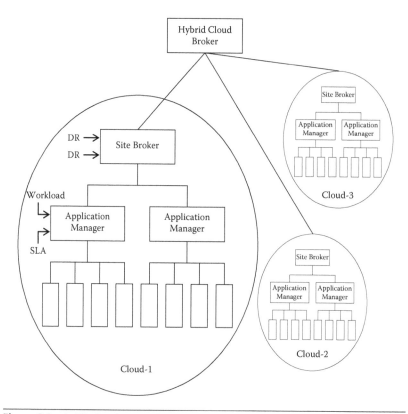

Figure 2.1   Architecture for managing application migration under performance constraints and in the presence of demand response signals.

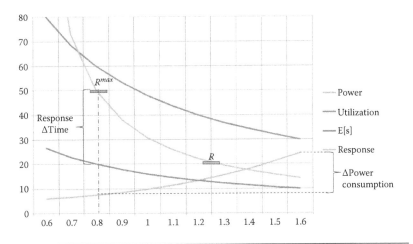

Figure 2.2 Relationship of response time, utilization, and power with operating frequency.

2.4.1 Application Manager

An application manager is responsible for managing an application locally at a cloud site and for trade-offs of the application's performance for savings in power consumption. Power performance trade-off possibilities include

1. Powering down a few servers and redistributing the excess workload, created as a consequence of shutting down these servers, onto the remaining servers[9,10] and
2. Operating each of the servers that host the application at a lower frequency and voltage using DVFS schemes.[11] Figure 2.2 shows the variation in power drawn, utilization, and response time as the operating frequency of the processor varies. The horizontal axis shows the processor frequency, from a maximum of 1.6 GHz to its minimal clock frequency of 0.6 GHz. As the frequency is reduced, the average service time per request increases. This causes an increase in the average processor utilization. High utilization then results in degraded average response time. The figure shows that, at the highest frequency of 1.6 GHz (and highest power consumption), the response time is approximately 25 ms. As the frequency of the process is reduced to 0.8 GHz, the utilization doubles. The response time reaches

its maximum acceptable level of 50 ms. This type of analysis allows us to examine the relationship of processor frequency to response time.

Accordingly, the role of the application manager is to determine for each application the optimal number of physical machines or the optimal value of the operating frequency and voltage so that the power consumption can be reduced to maximal levels without compromising the application's performance as guaranteed in the SLA. This in turn requires an application manager to address the question: How much savings in power consumption can be expected by allowing the application's response time to temporarily degrade to R_{max}? Another key issue that the application manager needs to resolve is to determine the time duration, as a fraction of the peak power grid load duration, for which threshold-level performance of the application is acceptable. The application manager communicates this information together with the power savings that it can achieve for the application that it manages to the site broker.

2.4.2 Site Broker

The site broker uses the information provided by the application managers and the DR signal available from the smart meter to sequence applications' execution for the duration in which the electric grid is experiencing peak load so that the total electricity consumed at a cloud site during this duration can be reduced to acceptable levels. In addition, the site broker is responsible for minimizing the number of applications that need to be moved to remote cloud sites to bring down the total electricity consumption to acceptable levels during such intervals. Moving applications to remote cloud sites should be the last priority and should be performed only when powering down servers and reducing the operating frequency of the servers does not solve the problem in its entirety.

2.4.3 Hybrid Cloud Broker

As discussed in the previous section, recent advancements in cloud computing make it possible to move "live" applications between physical servers located in different geographies.[5] The HCB leverages these new application mobility techniques to move applications identified by the site broker for migration to remote clouds. For each of these migrating applications, the HCB determines, from among the clouds managed by it, the cloud that is best suited as the new hosting environment. The HCB must take into consideration various constraints, such as the incompatibility constraints and the capacity constraints, while assigning an application to a particular cloud. The HCB initiates the migration of the applications identified by the site broker once suitable clouds that can host these applications have been identified.

2.5 Solutions

In the following sections, we formulate mathematical models and discuss solution procedures for the problems addressed by each of the three components discussed previously.

2.5.1 *Application Manager*

In this section, we establish a mathematical relationship between the response time of requests to the workload and the operating frequency of the processors. The derived system model assumes that an application i is hosted in a clustered environment and consists of a set of front-end servers responsible for accepting requests and another set of back-end servers responsible for processing the accepted requests. This assumption is not prohibitive as the architecture discussed in the previous section is extensible, and system models for various other scenarios can be easily incorporated as part of the application manager. We further assume that each of the back-end servers supports DVFS such that the operating frequency of the processors can vary at discrete intervals. Assume that f^{max} is the maximum frequency at which the servers can operate. Let μ_j^{max} be the service rate of the jth server when operating at f^{max}. The service rate μ_j of the server when operating at frequency f_j ($f_j < f^{max}$) then becomes $\mu_j = (\mu_j^{max} f_j)/f^{max}$. If the servers are homogeneous, μ_j^{max} is same for all the servers and is denoted using μ^{max}. The power consumption by a back-end server P_j can be modeled mathematically as $aj + b_j f_j^3$, where α_j and β_j are standard parameters obtained from regression tests on empirically collected data. The application manager needs to determine the operating frequency f_j of each server j and the number of active servers so that the aggregate power consumption is minimized, and the response time guarantees associated with the application operating at threshold SLA levels are met. Thus,

$$Min \sum_{j=1}^{N} X_j (\alpha_j + \beta_j f_j^3)$$

Subject to:

$$\sum_{j=1}^{N} X_j \lambda_j = \lambda_i \qquad (2.1)$$

$$R_j(\mu_j, \lambda_j) \le R^{max} \qquad (2.2)$$

According to the *M/M/1* queuing model $R_j = 1/(\mu_j - \lambda_j)$. Substituting $\mu_j = (\mu_j^{max} f_j)/f^{max}$ and $R_j = R^{max}$, we obtain

$$R^{max} = 1/\left((\mu_j^{max} f_j)/f^{max} - \lambda_j\right)$$

and reorganizing $f_j = (f^{max}/\mu_j^{max})(\lambda_j + 1/R^{max})$. On substituting the expression of f_j, the objective function becomes

$$\sum_{j=1}^{N} X_j \left(\alpha_j + \beta_j \left(\left(f^{max}/\mu_j^{max} \right)\left(\lambda_j + \left(1/R^{max} \right) \right) \right)^3 \right)$$

and is untenable for standard solvers. We therefore devise heuristic algorithms for solving the problem in a realistic time. Figure 2.3 shows the details of our algorithm.

Let λ_i represent the load experienced by the application and λ_j represent the load handled by the server j. Further, assume N is the total number of machines catering to the application load for ensuring a response time of R_{avg}. In other words, N is the upper bound for the number of servers to be used for operating the application at threshold SLA levels. In reality, not all of these servers may be used when the application operates at threshold SLA levels. The algorithm

1. Initialize $\lambda_i' = \lambda_i$ and a list of machines, $J = \{j \mid j = 1, \ldots, N\}$.
2. Select a machine j from the list J. Calculate

$$f_j' = \sqrt[3]{\frac{2\alpha_j}{\beta_j}}.$$

3. Set the frequency f_j of machine j as $\min(f^{max}, f_j')$.
4. Calculate the number of requests handled by machine j as

$$\lambda_j = \frac{f_j * \mu_j^{max}}{f^{max}} - \frac{1}{R^{max}}.$$

5. Calculate $\lambda_i' = \lambda_i' - \lambda_j$. If $\lambda_i' > 0$, remove j from the list of machines, that is, $J = J - \{j\}$. If J is nonempty, go to step 2.
6. If $\lambda_i' \leq 0$ and J is nonempty, all machines j in J can be powered down.
7. If $\lambda_i' > 0$ and J is empty, calculate

$$f' = \frac{f^{max}}{\mu^{max}}\left(\frac{\lambda_i'}{N} + \frac{1}{R^{max}}\right).$$

8. The operating frequency for all the servers j then becomes $f_j + f'$.

Figure 2.3 Outline of the provisioning algorithm for an application manager.

rests on the observation that whenever the difference between the terms $\beta_j f_j^3$ and α_j is greater than α_j, it is beneficial from the power consumption perspective to switch to a new machine (step 2). Thus, the frequency for a machine j is governed by the relationship $\beta_j f_j^3 - \alpha_j \geq \alpha_j$. Rearranging the equation results in $f_j' = \sqrt[3]{2\alpha_j / \beta_j}$. Since the operating frequency of all machines is limited by f^{ma}, the frequency f_j of machine j is set as the minimum of f^{ma} and $\sqrt[3]{2\alpha_j / \beta_j}$ (step 3). Rearranging $f_j = f^{max}/\mu_j^{max})(\lambda_j + (1/R^{max}))$ and setting f_j as $min(f^{max}, \sqrt[3]{2\alpha_j / \beta_j})$, we obtain the amount of load handled by machine j as $\lambda_j = (f_j * \mu_j^{max}/f^{max}) - (1/R^{max})$ (step 4). The amount of load handled by machine j is then subtracted from the total remaining load λ_i' still to be allocated. The remaining load λ_i' needs to be assigned to the rest of servers $J = J - \{j\}$ (step 5). In case the entire load has been assigned ($\lambda'_i \leq 0$), a machine that has not been allocated any load can be powered down (step 6). It is also possible that there is a fraction of the load that remains unassigned in spite of every machine receiving a portion of the load. In such cases, the unassigned load needs to be distributed among all the machines, and the operating frequency of the machines needs to be increased by $(f^{\max}/\mu^{\max})((\lambda'_i/N) + (1/R^{\max}))$ (step 7).

Next, we discuss the duration τ_i for which an application can afford to operate at threshold SLA levels. Assume T is the duration for which a peak electric grid load situation exists. Further, let us denote the period immediately following T by T'. Since an application needs to maintain an average response time of R^{avg} over $T + T'$, the following equation must hold true:

$$\frac{(\lambda_i \tau_i)R^{\max} + (\lambda_i(T - \tau_i))R^{avg} + (\hat{\lambda}_i T')R'}{(\lambda_i T) + (\hat{\lambda}_i T')} = R^{avg},$$

where λ_i is the load during the period T (T is also the period when a peak power grid load situation occurs), and $\hat{\lambda}_i$ is the forecasted load for the time period T' immediately following T. Our objective is to determine τ_i. However, the equation has an additional unknown variable R'. To determine R', we note that a large value of τ_i ($\tau_i < T$) is desirable even though τ_i need not exactly be equal to T. During T', the aim is to compensate the deviation from R^{avg} to $R^{\max}$, during T, by operating the application at maximum frequencies so that the response times are minimized. Thus, R' can be approximated as

$$\frac{1}{\mu^{\max} - \lambda_i/N}.$$

Substituting the value of R' in the previous equation results in

$$\tau_i = \frac{R^{avg}(\lambda_i T + \hat{\lambda}_i T') - \hat{\lambda}_i T' \dfrac{1}{\mu^{\max} - \lambda_i/N} - \lambda_i T R^{avg}}{\lambda_i (R^{\max} - R^{avg})}$$

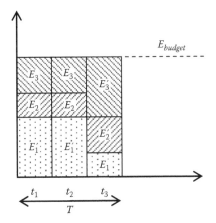

Figure 2.4   The optimal solution to the resource allocation problem addressed by the site broker.

2.5.2 Site Broker

In this section, we describe the resource management problem addressed by the site broker. The complexity of the problem can be observed from the example shown in Figure 2.4. Assume there are four applications, $i = 1, 2, 3, 4$, with power requirements 0.5, 0.3, 0.5, and 0.4, respectively, for operating at standard SLA levels. Further assume that for the duration for which the peak power grid situation exists, T is 3 units. Application 1 can afford to operate at threshold SLA performance levels for 1 time unit, and the power that it consumes is 0.2 units. Similarly, application 2 can operate at threshold levels for 2 time units and consumes 0.2 power units. Finally, applications 3 and 4 can operate at threshold SLA levels for 2 time units each and consume 0.3 and 0.35 power units, respectively. If the power budget is for 1 unit, it can be verified that the optimal solution is to move application 4 to a different cloud site. The sequencing of the remaining applications and their associated power consumption details are shown in Figure 2.4. All other configurations are suboptimal. Figure 2.5 shows one such suboptimal scheduling of applications. Thus, let:

P_i: Power consumed by application i if operating at reduced performance levels during peak power–grid load
P_i': Power consumed by application i if operating at normal performance levels.

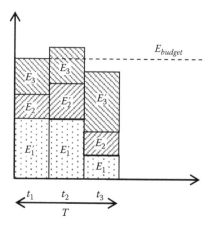

Figure 2.5   A suboptimal solution to the resource allocation problem addressed by the site broker.

P_{budget}: Average power budget during peak load

τ_i: Acceptable duration for executing application i at reduced performance levels during peak power–grid load

$$X_i = \begin{cases} 1 & \text{if application } i \text{ is migrated} \\ 0 & \text{o/w} \end{cases}$$

$$Y_i = \begin{cases} 1 & \text{if application } i \text{ is executing at reduced performance} \\ & \text{level at time-slot } t \\ 0 & \text{o/w} \end{cases}$$

T: Time duration for which peak power-grid load situation exists.

We divide T into N slots of duration δ each, $N = \lceil T/\delta \rceil$. We use the identifier t for these time slots. Then, $n_i = \lceil \tau_i/\delta \rceil$. Further, assume that E_i is the power consumed by an application i in one time slot, and E_{budget} is the average power consumed by all applications in one time slot. Then, $E_i = P_i/n_i$, $E'_i = P'_i/(N - n_i)$, and $E_{budget} = P_{budget}/N$. Mathematically, the problem that the site broker addresses can be formulated as

$$Min \sum_i X_i$$

Subject to:

$$\sum_{t=1}^{N} Y_{it} = n_i(1 - X_i) \qquad \forall i$$

$$Y_{it} \le (1 - X_i) \qquad \forall i,t$$

$$\sum_i E_i Y_{it} + \sum_i E_i'(1 - Y_{it}) \le E_{budget} \qquad \forall t$$

The computational time to determine an optimal solution to this problem increases exponentially with the problem size as the problem is nondeterministic polynomial (NP) hard. We therefore propose heuristics that can provide a sufficiently good solution in a reasonable time. We sort the applications in the decreasing order of $(E_i' - E_i) * n_i / E_i' * (N - n_i)$. The numerator indicates the total power savings generated by an application i when the power grid experiences peak load. This savings in power is due to the application operating at threshold SLA levels for a fraction of time within the period T and is an indicator of the benefits for retaining application i for execution on the current cloud site. The denominator indicates the nominal power consumed by an application during time period T when operating under standard SLA levels and is an indicator of the cost for retaining application i for execution in the current cloud site. Applications that do not qualify a certain user-defined threshold are candidates for migration to remote cloud sites. We represent the set of applications that are retained using the notation Γ. We propose a sequencing heuristic for applications that have been retained for identifying the time instances within the period T when an application should operate at threshold SLA levels. Assume the power consumption of each application i is represented using blocks of two sizes: i_1 and i_2. A block with size i_1 represents the application operating under standard SLA conditions. The block with size i_2 represents the application operating under threshold SLA conditions. The heuristic is motivated by our observation that since $i_1' > i_2'$ and $i_1 > i_2$, blocks with size i_1' are scheduled for execution together with blocks with size i_2, and blocks with size i_1 are scheduled for execution together with blocks with size i_2'. In addition, there could be blocks of size i_1' that need to be combined with blocks of size i_1. This results in three blocks of sizes $i_1' + i_2$, $i_1 + i_2'$ and $i_1' + i_1$ (or $i_2' + i_2$). Thus, at iteration l there would be $l + 1$ blocks of different sizes. If the size of any of the blocks exceeds the budget E_{budget}, the algorithm terminates. All remaining applications are considered candidates for cloud migration. It is to be noted that the heuristic selects applications in the decreasing order of $(E_i' - E_i) * n_i/E_i' * (N - n_i)$ for maximizing packing efficiency.

2.5.3 Hybrid Cloud Broker

The site broker communicates the details of the applications that need to be migrated to remote cloud sites to the HCB. It is assumed that each migrating application has at least one alternate cloud where it can be hosted. If there are multiple remote clouds that can host a migrating application, the criteria used by the HCB to select a particular cloud for hosting is that the degradation in the performance metric as a result of rehosting should be minimal. Let D_i be the amount of data associated with an application i that needs to be moved and $V_{kk'}$ be the network traffic between clouds k and k'. Let $\bar{k} = \arg\min_{k'}(D_i/V_{kk'})$. It is then decided to move application i to cloud $\bar{k}$ from its current hosting cloud k. The HCB then initiates contact with the site broker of the remote cloud site. The site broker communicates resource allocation details to the provisioning manager, which then provisions sufficient computing resources on the physical machines identified by the site broker. The HCB can then initiate the actual physical movement of the application to the machine identified by the site broker previously. Figure 2.6 shows that the total energy consumption by applications executing at cloud A during the peak electric grid load period is high (indicated by the bottom dot in cloud A). When a DR signal is received by the smart meter installed at cloud A, the site broker at cloud A identifies a set of applications for which the operating frequency can be lowered for a fraction of the peak duration and another set of applications requiring migration to cloud B as shown in Figure 2.7. Figure 2.7 shows the reduction in energy consumption at cloud A after a set of applications is migrated from cloud A (three dots) to cloud B (indicated by the top two dots). The HCB is responsible for orchestrating all intercloud movements of the VMs.

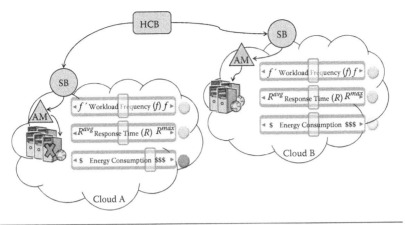

Figure 2.6 Scenario before application migration when a demand response signal is received by a smart meter at cloud A. AM = application manager; SB = site broker.

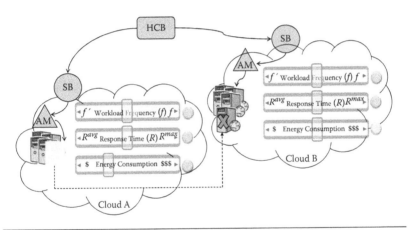

Figure 2.7 Scenario after application migration after a demand response signal is received by a smart meter at cloud A.

2.6 Smart Meters and Smart Loads

The term *smart grid* is used in varied contexts: For some, the smart grid means a superhighway for large-capacity transmission over significant geographic distances; for others, it is a system that can integrate small-scale renewable generation sources; still others see it as a widely available source network for charging electric cars. Perhaps it will turn out to be some or all of these things when fully evolved. In the context of the data center or cloud, we can shed much of this ambiguity and need only consider specifically the functionality required to enable the capability to monitor real-time power consumption and costs.

2.6.1 *The Data Center Smart Grid*

We are interested in determining the real-time marginal cost of power for an incremental load. The two key parameters we require the smart grid or smart meter to provide information about is the real (or near-real-time) power load for the data center, typically measured in kilowatts, and current price, usually quoted per kilowatt-hour. So, in this respect, the requirements for calculating marginal power costs of the data center in which one or more clouds reside are rather modest. Existing data center sites do not have smart meters capable of providing the data we need. Currently, these sites have a standard power meter that is based on electromechanical induction and needs to be read at the end of a billing cycle by the electric utility. Some utilities are using newer meters that allow these data to be collected remotely, but many facilities pre-date this effort. Electricity meters continuously measure the instantaneous voltage (volts) and current (amperes) and find the product of these to give instantaneous electrical power (watts). Watts are then integrated over time to give energy used (joules, kilowatt-hours, etc.).

The cost of a kilowatt-hour of electricity is variable depending on many factors; typically, these specific data are not provided by existing meters. In fact, a utility does not typically even calculate this cost until the end of a billing cycle. Most utilities determine the cost for electricity on a tiered rate structure. This results in a cost model that has discontinuous step function increases in cost. A baseline number of kilowatt-hours is provided at a certain price; when cumulative use for a billing interval exceeds this baseline amount, the consumer is then charged at an increased price for the next tier for another fixed amount of kilowatt-hours, and so on. While the cost of electricity is variable and is data not typically provided by the utility nowhere close to real time, we need only note that it is readily calculable based only on knowledge of the start of the billing cycle, the tier rates, the tier amounts, and total metered usage to a point in time.

While we lack the smart grid of the future that includes smart meters that provide real-time price signals and current electricity usage, this functionality can be implemented in a data center with relative ease and at a moderate cost. What is required is metering independent of the utility. This could be implemented with any number of digital power meters available on the market. This meter must provide remote monitoring so that the data can be polled in real or near-real time by a computer system. We envision that these data would be available to the smart meter (SM) via polling. This independent power meter need not necessarily be shunt based and thus required to be in line with the existing utility meter; there are types of meters that measure the power usage by inductive magnets that need only encircle the existing physical wires that deliver the main power, minimizing installation costs. The other requirement is that the SM be able to poll the meter via a communication protocol that is a common feature of existing digital power meters. The SM will know at any point in time the billing period and the cumulative power consumption and thus be able to accurately estimate the cost of electricity at that point in time.

The cost of continuing operation of a load in the data center or of shedding that load is then easily calculated and provides the cost input that the SM together with the HCB uses to make the decision to move local loads to remote sites.

2.6.2 Smart Appliances in the Data Center

We should recognize that moving a VM from a local physical machine to a remote physical machine does result in a shedding of some load; however, the local physical machine that is still powered on can be seen to be consuming electricity nonproductively. Simply put, the maximum load is shed when the physical machine not hosting any VMs is powered off. In fact, the economic premise underpinning the smart grid is that a real-time price signal, assuming resource consumers are economically rational, will necessarily result in more efficient use. A tacit assumption here is that consumers are able to respond to the price signal. It is from this premise and assumption that smart grid proponents argue that the market will self-evidently provide smart appliances that enable load shedding by time-of-day delay of operation or power-saving mode settings.

With respect to the data center, the immense existing capital equipment will not be replaced with next year's smart-grid-friendly models as a matter of course. So, we will have to deal with existing infrastructures. However, existing servers typically do have power-saving modes that can be invoked via network-initiated commands. When hibernating or in power-saving modes, typically servers can be configured to respond to wake-on-LAN (WOL) signals. When VMs and their loads are shed from physical machines, when possible the SM should initiate the power-saving modes to maximize the load that is shed. In this case, the load shed is the difference between the power load for the physical machine when the VMs are hosted and when the physical machine is in power hibernation mode.

It is possible to completely maximize the load shed by powering off the physical machine. While this command can be initiated with commands sent via a network connection to the physical machine, the main problem here is that typically in the off state a machine cannot be powered on remotely and usually requires manually pressing the power on button. One approach to resolve this issue is, in effect, to retrofit an existing server, equipment cabinet, or particular circuit. In such an implementation, one need only build a black box device that is in series to the power provided to the equipment under consideration. The minimal functionality of this black box provides mechanical or solid-state relay control of the input power source. If the on/off state of this relay can be controlled via network or other protocols, then the SM can power down the physical machine on commands. Existing servers can be easily configured to boot on the application or restoration of input power.

Existing network-controlled power busses and even commercial power meters provide exactly this functionality. Such devices can easily be placed in series with existing power inputs for the physical machine, equipment cabinet, or power circuit. The advantage of using a power meter with this relay functionality is that these meters typically have network-enabled communications. Thus, they also provide power consumption telemetry at this granular level and enable the HCB to exactly quantify how much load will be shed since the load when VMs are hosted can be measured; with this method, they know that the complete load will be shed when the system is powered down. This would be an SM-initiated network safe shutdown command to the host prior to an SM-initiated power down of the power source. Likewise, the SM can initiate the power on of a physical server by restoring the power source to the equipment it is controlling with this networked controlled relay.

So, while the smart grid and smart appliances may soon be realized in the not-too-distant future, the functionality we need the smart grid to provide the SM and the HCB is the ability to move resources and then completely shed the load associated with the physical machines that no longer host these resources by implementing existing off-the-shelf devices. It reduces to a simple cost-benefit analysis to justify the associated capital expenditures. The costs include the hardware to independently meter the site power source to determine current power costs that can be polled by the SM and HCB. Minimally, the SM and the HCB must be able to communicate with the physical servers that host the VMs to provide commands to invoke power-saving modes. In addition, hardware may be cost-benefit justified to provide the network controlled relay for input power at a server, cabinet, or circuit granularity. The communication protocol between the SM and the HCB and this hardware can be TCP/IP based, infrared (IR), radio frequency (RF), or perhaps even use the existing power line infrastructure using ZigBee or similar protocols.

2.7 Conclusions

The chapter described ways to reduce electricity usage within data centers during peak power grid load situations by identifying suitable applications whose performance can be traded off for short durations for savings in electricity consumption. Reducing electricity consumption during peak power grid loads is important for data centers as a significant portion of the electricity cost incurred by data centers could be due to the electricity consumed during peak power grid load-occurring situations. The chapter described a solution architecture and discussed analytical formulations together with heuristic schemes for minimizing the electricity consumed during peak power grid situations. The approach leverages recent technical developments in cloud computing that make it possible to move live applications across the wide-area network.

References

1. J. Choi, S. Govindan, B. Urgaonkar, and A. Sivasubramaniam, Profiling, prediction, and capping of power consumption in consolidated environments, In *Proceedings of 16th IEEE International Symposium on Modeling, Analysis and Simulation of Computers and Telecommunication Systems (MASCOTS)* (2008).

2. D. Kusic, J. Kephart, J. Hanson, N. Kandasamy, and G. Jiang, Power and performance management of virtualized computing via lookahead control, In *Proceedings of 5th International Conference on Autonomic Computing (ICAC)* (2008).

3. J. Kephart, H. Chan, R. Das, D. Levine, G. Tesauro, F. Rawson, and C. Lefurgy, Coordinating multiple autonomic managers to achieve specified power-performance tradeoffs, In *Proceedings of 4th International Conference on Autonomic Computing (ICAC)* (2007).

4. G. Tesauro, D. Chess, W. Walsh, R. Das, A. Segal, I. Whalley, J. Kephart, and S. White, A multi-agent systems approach to autonomic computing, In *Proceedings of 3rd International Joint Conference on Autonomous Agents and Multi-agent Systems (AAMAS)* (2004).

5. R. Bradford, E. Kotsovinos, A. Feldmann, and H. Schiöberg, Live wide-area migration of virtual machines including local persistent state, In *Proceedings of the 3rd International ACM/Usenix Conference on Virtual Execution Environments* (2007).

6. S. Osman et al. The design and implementation of Zap: A system for migrating computing environments, In *Proceedings of the 5th Symposium on Operating Systems Design and Implementation*, December 2002.

7. Jacob G. Hansen and Eric Jul. Self-migration of operating systems, In *Proceedings of the 11th ACM SIGOPS European Workshop* (EW 2004), pages 126-130, 2004.

8. M. Kozuch and M. Satyanarayanan, Internet Suspend/Resume. In WMCSA '02 *Proceedings of the Fourth IEEE Workshop on Mobile Computing Systems and Applications* (2002).

9. A. Gandhi, M. Harchol-Balter, R. Das, and C. Lefurgy, Optimal power allocation in server farms, In *Proceedings of 11th International Joint Conference on Measurement and Modeling of Computer Systems* (2009).

10. M. Steinder, I. Whalley, J. Hanson, and J. Kephart, Coordinated management of power usage and runtime performance, In *Proceedings of Network Operations and Management Symposium (NOMS)* (2008).

11. E. Elnozahy, M. Kistler, and R. Rajamony, Energy-efficient server clusters, In *Proceedings of 2nd Workshop on Power-Aware Computing Systems* (2002).

Bibliography

J. Chase and R. Doyle, Balance of power: Energy management for server clusters, 2001. http://www.cs.duke.edu/ari/publications/publications.html

Intel Corporation, *Enhanced Intel® SpeedStep® Technology for the Intel® Pentium® M Processor*, white paper, March 2004.

S. Kiliccote, M. Piette, G. Wikler, J. Prijyanonda, and A. Chiu, Installation and commissioning automated demand response systems, In *Proceedings of 16th National Conference on Building Commissioning* (2008).

D. Niyato, S. Chaisiri, and L. Sung, Optimal power management for server farm to support green computing, In *Proceedings of 9th IEEE/ACM International Symposium on Cluster Computing and the Grid* (2009).

A. Qureshi, R. Weber, H. Balakrishnan, J. Guttag, and B. Maggs, Cutting the electric bill for Internet-scale systems, In *Proceedings of the ACM SIGCOMM Conference on Data Communication (SIGCOMM '09)* (2009).

X. Ruibin, Z. Dakai, R. Cosmin, M. Rami, and M. Daniel, Energy-efficient policies for embedded clusters determine the number of active nodes, In *Proceedings of ACM SIGPLAN/SIGBED Conference on Languages, Compilers, and Tools for Embedded Systems* (2005).

3

DISTRIBUTED OPPORTUNISTIC SCHEDULING FOR BUILDING LOAD CONTROL

PEIZHONG YI, XIHUA DONG, ABIODUN IWAYEMI, AND CHI ZHOU

Contents

The smart grid adds intelligence and bidirectional communication capabilities to today's power grid, enabling utilities to provide real-time pricing (RTP) information to their customers via smart meters. This facilitates customers' participation in demand response programs to reduce peak electricity demand. In this chapter, we provide a novel distributed opportunistic scheduling scheme based on an optimal stopping rule that aims to minimize the expenditure of electricity while satisfying

85

customers' time requirements. The proposed scheduling scheme can be implemented in either centralized or distributed mode; constraint of a power line's total power consumption is also considered in the system model. Simulation results show it can dramatically reduce the electricity bill and minimize peak loads.

3.1 Introduction

The smart grid is an intelligent power generation, distribution, and control system equipped with two-way communication. It facilitates many services, including integration of renewable energy sources, real-time pricing (RTP) to consumers, demand response (DR) programs involving residential and commercial customers, and rapid outage detection.

According to a report from the U.S. Department of Energy (DoE), buildings consume 72% of all electrical energy.[1] Therefore, the ability of a building automation system (BAS) to communicate and coordinate with the power grid has tremendous potential to reduce the peak in response to pricing and demand reduction signals by utilizing smart meters located within customer sites. These devices provide customers and utilities real-time power consumption data and RTP information. The automation system facilitates this information for monitoring and controlling building loads and home appliances by an intelligent energy management algorithm.

The DR algorithm plays a key role in saving energy by the process of collecting, monitoring, controlling, and conserving energy in a building. It enables people to reduce costs, carbon emissions, and risk of increased price or supply shortages. Typically, this involves four steps: (1) metering and collection of the data of energy consumption and real-time price; (2) finding appliance shift opportunities and estimating energy saving; (3) monitoring the appliance to target the opportunities to save energy; and (4) tracking the progress by analyzing your meter data to see an effect. From this, we can see that information is the most important factor for estimation and planning. However, the day-ahead price cannot always match the real-time price due to some factors, such as those shown in Figure 3.1 (the data were collected July 11 to 15, 2011).

The weather can have a big impact on the wholesale real-time price of electricity, particularly during the summer and winter. There also can be unexpected and brief price spikes if multiple power plants have

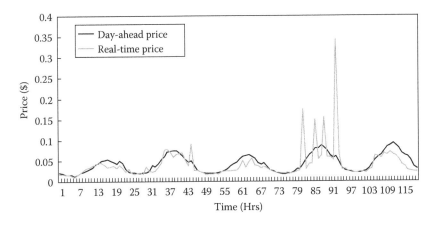

Figure 3.1 Day-ahead price versus real-time (Ameren Energy Illinois: July 11, 2011 to July 15, 2011).

technical or mechanical problems at the same time or if there are problems in parts of the regional transmission network used to transport electricity from the power plant to the distribution system.

In this chapter, we adopt a novel distributed opportunistic scheduling scheme based on the optimal stopping rule. The objective is to minimize the energy consumption at the peak time while satisfying the power and timing requirements of each utility. In comparison with a traditional centralized scheduling scheme, devices can adjust their service time and mode adaptively according to the real-time price without complicated computation. We show that the optimal scheduling scheme is a pure threshold policy; that is, each utility can be turned on when the electricity price is lower than a threshold value.

3.2 Demand Response

3.2.1 Power Pricing

Currently, the majority of residential customers are charged flat or two-tiered (peak and off-peak) electricity rates based on average electricity generation costs. The implication is that retail prices do not accurately reflect the actual cost of generating electricity at any given time. This results in inefficient investment in generation and transmission capacity and higher retail prices.[2] Due to the deficiencies of this scheme, a variety of pricing schemes has been introduced to more accurately pass on the true cost of electricity to retail customers. They include RTP, time-of-use (TOU) pricing, day-ahead pricing (DAP),

Table 3.1   DR Modes (Ameren Energy Illinois: January 17–21, 2011)

PRICE-BASED OPTIONS

Time of use (TOU): A rate with different unit prices for usage during different blocks of time, usually defined for a 24-h day. TOU rates reflect the average cost of generating and delivering power during those time periods.

Real-time pricing (RTP): A rate in which the price for electricity typically fluctuates hourly, reflecting changes in the wholesale price of electricity. Customers are typically notified of RTP prices on a day-ahead or hour-ahead basis.

Critical peak pricing (CPP): CPP rates are a hybrid of the TOU and RTP designs. The basic rate structure is TOU. However, provision is made for replacing the normal peak price with a much higher CPP event price under specified trigger conditions (e.g., when system reliability is compromised or supply prices are very high).

INCENTIVE-BASED PROGRAMS

Direct load control: A program by which the program operator remotely shuts down or cycles a customer's electrical equipment (e.g., air conditioner, water heater) on short notice. Direct load control programs are primarily offered to residential or small commercial customers.

Interruptible/curtailable (I/C) service: Curtailment options integrated into retail tariffs that provide a rate discount or bill credit for agreeing to reduce load during system contingencies. Penalties maybe assessed for failure to curtail. Interruptible programs have traditionally been offered only to the largest industrial (or commercial) customers.

Demand bidding/buyback programs: Customers offer bids to curtail based on wholesale electricity market prices or an equivalent. Mainly offered to large customers (e.g., those using 1 MW and more).

Emergency demand response programs: Programs that provide incentive payments to customers for load reductions during periods when reserve shortfalls arise.

Capacity market programs: Customers offer load curtailments as system capacity to replace conventional generation or delivery resources. Customers typically receive day-of notice of events. Incentives usually consist of up-front reservation payments, and face penalties for failure to curtail when called on to do so.

Source: Office of Electricity Delivery and Energy Reliability, "Benefits of Demand Response in Electricity Markets and Recommendations for Achieving Them," U.S Department of Energy, 2006.

and critical peak pricing (CPP). An explanation of the pricing terms and various DR schemes are provided in Table 3.1.

3.2.2 Demand Response

Demand response is defined as "changes in electric usage by end-use customers from their normal consumption patterns in response to changes in the price of electricity over time, or to incentive payments designed to induce lower electricity use at times of high wholesale market prices or when system reliability is jeopardized."[2] DR programs fall into two categories: price-based and incentive DR programs.

Price-based DR methods include RTP, TOU, and CPP schemes. The retail price of electricity varies on an hourly basis to reflect real-time wholesale electricity costs, and residential customers save money by either reducing their energy consumption during peak periods or shifting it to off-peak periods. This DR scheme requires the availability of smart meters and an advanced metering infrastructure (AMI) to facilitate the communication of RTP to customers.

Incentive-based DR schemes pay customers to reduce their electricity consumption when the price of electricity is high or when the stability of the power grid is under threat due to excessive demand. These schemes typically involve the installation of a switch that enables the utility to cycle residential air conditioners or water heaters when prices (or system loads) are high. This scheme is also termed *direct load control*.

3.2.3 DR Benefits

DR adoption promises benefits in several areas, including lower retail prices due to reduction in the need for expensive peaking power plants (i.e., power plants used only when there is high or peak power demand), increased grid reliability due to the avoidance of power outages, and reduction in the need for new generation capacity due to reduced demand.

3.2.4 DR Guidelines

To guarantee widespread adoption and fulfill the potential of DR, residential DR infrastructure must have the following features: automation; scalability to large areas; control of intelligent and legacy (or dumb) home appliances; the integration of renewable energy sources such as solar cell arrays and plug-in hybrid electric vehicles (PHEVs); and avoidance of the creation of rebound peaks that can result from shifting electricity usage to off-peak periods.[3]

3.3 Optimal Stopping Rule

The optimal stopping rule is the perfect tool for DR in a BAS. The method addresses the problem of determining the best time to take an action on an observed sequence of random variables to maximize expected rewards or minimize expected costs.[4] Specifically, let $(X_1; X_2; \ldots)$ denote a random process whose joint distribution is assumed known, and $(y_0; y_1(x_1); y_2(x_1; x_2); \ldots; y_\infty(x_1; x_2; \ldots))$ denote a sequence of real-valued reward functions. We need to choose a stopping time N that satisfies $\{N = n\} \in F_n$ where F_n is the σ algebra generated by $(X_j; j < n)$ to maximize or minimize the expected return $E[Y_N]$. It has been used effectively in statistics, economics, mathematics, finance, and networks. In our scenario, an action means an electricity user or appliance starts to run; the observation is the electricity price, and the objective is to minimize cost or maximize profit.

One of the most popular examples of an optimal stopping problem is the "secretary problem": A boss needs to select a perfect secretary from N applicants, for which N is known. All applicants can be ranked from best to worst without ties. They will be interviewed in a random order, and the boss has no information about the candidates before the interview. After each interview, the boss must make a decision to either accept the candidate on the spot or lose the chance forever. Once the applicant is rejected, he or she cannot be recalled. How can we guarantee the boss chooses the best secretary?

Intuitively, if we reject the first 50 percent of all applicants and choose the first applicant with a score better than all those already observed (and rejected), then we have greater than a 25% probability of winning. With the optimal stopping rule, we can find that if we do not select from the first 37% of candidates and choose the next interviewee whose rank is higher than the previous highest one, the winning probability increases to 36%. If we consider choosing the first or second best as winning, then a 57.4% winning probability can be achieved.

3.4 Problem Formulation

In this work, we use the optimal stopping rule to model your problem; more details can be found in Yi et al.[5] A *task* is denoted here as the minimum unit of an electricity user's work, which can be dish washing, PHEV charging, operating electrical machines, and so on. The electricity price process is modeled by a random process $P(t)$, and the time is divided into slots with length τ. We assume that, once

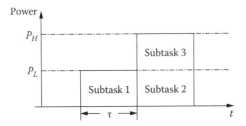

Figure 3.2 Task decomposition.

started, a task can be completed within one time slot. For some tasks, this assumption is valid (e.g., dish washing and clothes washing) since they can be generally completed within a short period. However, some tasks may require a much longer time and may have time-varying power. In this case, we can decompose this task to multiple subtasks so that each subtask can be completed within one time slot, and more importantly, we can schedule a task by simply starting or postponing a subtask. For example, consider a task that requires two time slots and has power P_L in the first time slot and P_H in the second time slot as shown in Figure 3.2.

We decompose the task to three subtasks: subtask 1, subtask 2, and subtask 3 with power P_L, P_L, and $P_H - P_L$, respectively. Another important reason for task decomposition is that some users (e.g., refrigerator) require minimum standby power levels (e.g., P_L in Figure 3.2); thus, part of the power consumption cannot be scheduled. By the task decomposition method, we can allocate a portion of the power consumption and schedule this part (e.g., task 3) more flexibly. Different from other scheduling schemes proposed by previous works (e.g., Mohsenian-Rad and Leon-Garcia[6]), which schedule users by changing the power level, our proposed scheduling policy is binary; that is, we need to decide whether to start or postpone a task in a time slot. Thus, it is the task decomposition method that makes our scheduling policy feasible.

Based on this discussion, in the remainder of this chapter, we always make the following assumption:

Assumption 1: Once started, a task can be completed within one time slot; that is, during the service time of a task, the electricity price is constant.

On the other hand, task decomposition may also result in dependence between tasks. For example, some tasks may require executing consecutively, which may complicate the scheduling problem. However, in this initial work, this type of dependence is not considered and is left to future work.

We sometimes use the expression "scheduling of tasks" and sometimes use the expression "scheduling of electricity users." It is obvious they have the same function. Moreover, although each electricity user may have multiple tasks, for simplicity we assume different tasks belong to different users. Thus, in the reminder of this chapter, we only talk about the scheduling of electricity users.

We consider a power system in which the tasks of electricity users arrive randomly. Recall that the electricity price process is modeled by a random process $P(t)$, and the time is divided into slots with equal length τ. Based on assumption 1, we also assume that once a user starts to operate, the electricity price is constant during its service time. Since we use a discrete-time model, we assume all arrivals take place at the beginning of a time slot. We assume the number of arrivals in a time slot is Poisson distributed with mean $\lambda \times \tau$, where λ denotes the average arrival rate. Specifically, let S_t denote the set of arrivals in the tth time slot; we have

$$\Pr\left(\left|S_t\right| = k\right) = \frac{(\lambda\tau)^k}{k!}e^{-\lambda\tau}, \; k = 0, \; 1, \; \ldots \tag{3.1}$$

where $|\cdot|$ denotes the cardinality of a set. It is well known that Poisson distribution is a good model of many service arrival processes; other examples include radioactive decay of atoms, telephone calls arriving at a switchboard, and others.

Now, we consider an arbitrary electricity user i. Let g_i denote its electricity consumption during its service time and A_i denote its arrival time. Let N denote a scheduling policy that determines when to start a user and N_i denote the corresponding scheduled operation time of user i. Then, the waiting time of user i is

$$W_i = N(i) - A_i \tag{3.2}$$

For a given user i, there are two sources of costs: (1) cost due to purchasing electricity, denoted by C_i^p, and (2) cost due to the waiting time, denoted by C_i^w. Then, it is easy to see

$$C_i^p = g_i \times P(N(i)) \tag{3.3}$$

Moreover, in this initial work, we assume the cost due to waiting is a linear function of the waiting time, that is,

$$C_i^w = \mu_i \times \tau \times W_i \qquad (3.4)$$

where μ_i is a positive constant and is referred to as the time factor of user i in this work. The total cost of user i is $C_i^p + C_i^w$. We are interested in the long-term average of the total cost, which is given as

$$C(N) = \lim_{M \to \infty} \frac{1}{M} \sum_{t=1}^{M} \left(E \sum_{i \in S_t} \left(C_i^p + C_i^w \right) \right) \qquad (3.5)$$

We also assume that there is a constraint on the total power consumption; that is, at any time t, the total power consumption satisfies

$$\sum_{l=1}^{t} \sum_{i \in S_l} g_i \times \delta(N(i), t) \le Q \qquad (3.6)$$

where $\delta(N(i), t) = 1$, if $N(i) = t$ and $\delta(N(i), t) = 0$ otherwise, and Q is the power constraint. Notice that in the inequality, we sum the time from 1 to t because all arrivals before time t may be scheduled to operate at time t. One of the major reasons for us to consider a power constraint is that with opportunistic scheduling, many users may operate around the off-peak time and thus produce a peak power requirement, which may be a challenge for the facilities.

Based on this discussion, we are now ready to formulate the cost-minimizing opportunistic scheduling problem as the following optimization problem:

The Cost-Minimization Problem

$$\min_{N \in C} \lim_{M \to \infty} \frac{1}{M} \sum_{t=1}^{M} \left(E \sum_{i \in S_t} \left(C_i^p + C_i^w \right) \right) \qquad (3.7)$$

$$s.t. \ \sum_{l=1}^{t} \sum_{i \in S_l} g_i \times \delta(N(i), t) \le Q \qquad (3.8)$$

where C is the class of scheduling policies.

Remark 1: For convenience, we have assumed the number of arrivals in a time slot has a Poisson distribution. However, as we show in the remainder, our major results do not depend on the arrival distribution. So, our work can be directly extended to address other arrival processes.

Remark 2: Problem (3.7) is a general form of the infinite-user, infinite-horizon cost minimization. The finite-user problem (i.e., scheduling of a finite number of electricity users) or finite-horizon problem (i.e., users have deadlines) can be formulated in a similar way. Besides, we show that our solution in Section 3.3 can be directly applied to the finite-user case (infinite-time horizon). For the finite-horizon problem, our solution can also be applied by making a slight modification (the finite-horizon optimal stopping problem can be solved by the dynamic programming approach[4]). In Section 3.4, we provide simulation results for different cases.

Remark 3: If we are interested in maximizing the rate of return instead of total cost, we can also formulate the scheduling problem as a dual profit-maximization problem. In our previous work,[7] we have shown that for the single-user case, the cost-minimization problem and the profit-maximization problem are essentially equivalent.

3.5 Simulation and Result

In this section, we apply our optimal stopping method to actual RTP data from the Ameren Web site (https://www2.ameren.com/RetailEnergy/realtimeprices.aspx, July 14–23, 2011) to evaluate the performance of our residential scheduling scheme. We take a clothes dryer as our simulation parameter. Normally, the running time of a clothes dryer is 0.75 h. The average power in a running cycle is 3 kW, and peak energy in a cycle is 6 kW. We can see from Figure 3.3 that the typical clothes dryer use time is during the day, and the peak hour occurs around 11 a.m., which is also the peak daily electricity price. Due to its high peak energy in the cycle and short service time, it offers significant opportunities for shifting peak electricity usage.

Figures 3.4 and 3.5 show the performance of our proposed scheduling scheme using the optimal stopping rule (OSR) and no optimal

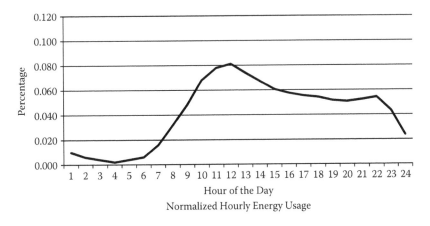

Figure 3.3   Normalized hourly residential energy usage.

stopping rule (NOSR). It can dramatically reduce the cost of electricity, with cost savings up to 50%. Average waiting time is acceptable. The time factor reflects the customer's time requirement for this appliance. A larger time factor means residents are more sensitive with the appliance, and it is less flexible to schedule. Small time factors are more suitable for DR and can save more electricity costs.

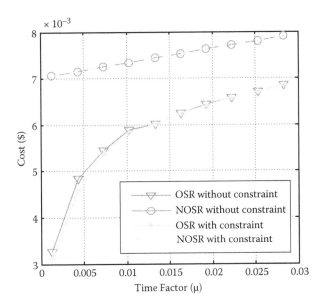

Figure 3.4   Clothes dryer time factor versus cost. OSR, optimal stopping rule; NOSR, no optimal stopping rule.

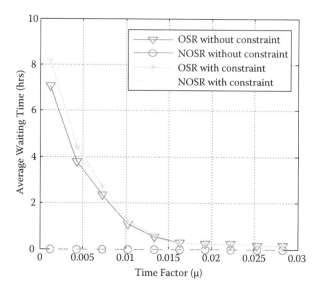

Figure 3.5 Clothes dryer time factor versus average waiting time.

3.6 Discussion

3.6.1 Modeling of Price Signals

In our preliminary work,[7] the price signal was modeled as an independent and identically distributed (i.i.d.) random process, which is not
realistic. We propose several other models for future study. The optimal stopping solution for non-i.i.d. price signals is difficult to obtain.

3.6.1.1 Random Modeling of Price Signals   Though modeled as random
processes in this proposal, the price signals often contain deterministic
components, such as seasonal components (period) and a trend. These
deterministic components can be utilized to estimate the electricity
price. Thus, a general electricity price process can be decomposed as

$$P(t) = m(t) + s(t) + w(t) \tag{3.9}$$

where $s(t)$ is a periodic function, $m(t)$ is a deterministic function, and
$w(t)$ is a zero-mean random process. This decomposition is exactly the
classical decomposition model of time series.[9]

   Now, we discuss these three components in detail. The seasonal
component $s(t)$ describes the short-term variation of price signals. For
example, there are peaks at the high-demand afternoons and troughs at
the low-demand nighttime hours. In this case, (t) has a period of 24 h.

The trend component $m(t)$ describes the long-term variation of price signals. For example, due to high energy demand for heating or air conditioning, the electricity price may be higher in winter and summer, respectively. The third component $w(t)$ describes the randomness of price signals. In our preliminary work,[7] we assumed that the $P(t) = w(t)$ is an i.i.d. random process. However, since the electricity process has memory in general (i.e., the electricity price does not change too much over two consecutive slots), a more realistic candidate for the random component $w(t)$ is the finite-state Markov chain (FSMC) model. A counterpart in wireless communication is the FSMC modeling of fading channels.[10] Building the model of price signals will be our first task. Many approaches, such as Wiener filtering, curve fitting (e.g., polynomial fitting), will be adopted in our future analysis.

3.6.1.2 Usage-Dependent Electricity Price   In the last section, we discussed random modeling of price signals. The price signal is assumed to be independent of customer usage. In some scenarios, the electricity price can be expressed as a deterministic function of customer usage. An example is the electricity market with an inclining block rate (IBR) pricing scheme, which has been widely adopted in the pricing tariffs by many utility companies since the 1980s (e.g., Southern California Edison, San Diego Gas and Electric, and Pacific Gas and Electric). In IBR pricing, the marginal price increases by the total quantity consumed.[11] That is, beyond a certain threshold in the total monthly/daily/hourly load, the electricity price will increase to a higher value. The IBR pricing scheme can stimulate the customers to distribute their load at different times of day to avoid paying for electricity at higher rates. Moreover, IBR pricing helps in load balancing and reducing the peak-to-average ratio (PAR).[6] With IBR pricing, the price signal can be expressed as

$$P(t) = f(u(t)); \qquad (3.10)$$

where $f(\cdot)$ is a deterministic step function, and $u(t)$ is the total energy consumption in the tth time slot. The random model and the usage-dependent model can be combined to provide a more general model of price signals. In this case, the price signal can be expressed by

$$P(t) = f(u(t); t) + w(t) \qquad (3.11)$$

3.6.2 Fairness

In the studies discussed, the objective of optimal scheduling is to minimize the total cost of multiple appliances. In some scenarios, fairness must be taken into consideration. For example, we consider a system with multiple independent customers; each user has a personal interest. Under fairness constraints, the objective of scheduling becomes minimizing the total utility function. There are two types of fairness can be considered, max-min fairness and proportional fairness. Now, we consider proportional fairness. The utility function can take the following form:[13]

$$U_k = \begin{cases} \log r, & \text{if } k = 1 \\ (1-k)^{-1} r^{1-k}, & \text{if } k \geq 0, \ k-1. \end{cases} \tag{3.12}$$

By solving the min-cost problem in (3.7) with C_i replaced by $U_k(C_i)$, we can get the optimal scheduling scheme under the proportional fairness constraint.

3.7 Conclusion

In this work, we presented our optimal stopping-based scheduling framework for building DR. The real-time price was modeled as random variables, and each appliance was assigned a different time factor. Our scheme automatically determined the best time to run the appliance to save the expenditure of electricity without waiting a long time. Results showed that our user-friendly scheduling scheme can reduce domestic energy consumption with minimal user intrusion while mitigating peak rebound. Future work includes investigation of price signal modeling, fairness, and incorporation of PHEVs into our framework.

Acknowledgment

This work is funded by the US Department of Energy under grant DE-FC26-08NT02875.

References

1. U.S. Department of Commerce, Integration of Building Control Systems/ Smart Utility Grid Project. http://www.nist.gov/el/highperformance_ buildings/intelligence/smartgrid.cfm (accessed April 24, 2011).
2. Office of Electricity Delivery and Energy Reliability, *Benefits of Demand Response in Electricity Markets and Recommendations for Achieving Them.* Washington, DC: U.S Department of Energy, 2006.
3. M. LeMay, R. Nelli, G. Gross, and C. A. Gunter, An integrated architecture for demand response communications and control, *Proceedings of the 41st Annual Hawaii International Conference on System Sciences*, January 2008, Waikoloa, Big Island, Hawaii, p. 174.
4. T. Ferguson, Optimal stopping and applications. *Electronic notes*, www. math.ucla.edu/~tom/stopping/Contents.html
5. P.Yi, X. Dong, A. Iwayemi, and C. Zhou, Real-time opportunistic scheduling for residential demand response, accepted by *IEEE Transactions on Smart Grid*, 4(1):227–234 (2013).
6. A.-H. Mohsenian-Rad and A. Leon-Garcia, Optimal residential load control with price prediction in real-time electricity pricing environments, *IEEE Transactions on Smart Grid*, 1(2):120–133 (2010).
7. P. Yi, X. Dong, and C. Zhou, Optimal energy management for smart grid systems—An optimal stopping rule approach, IFAC World Congress Invited Session on Smart Grids, August 2011, Milan, Italy.
8. A. Iwayemi, P. Yi, X. Dong, C. Zhou, Knowing when to act: An optimal stopping method for smart grid demand response. *IEEE Network* 25(5): 44-49 (2011).
9. P. J. Brockwell and R. A. Davis, *Introduction to Time Series and Forecasting*, 2nd ed. New York: Springer, 2002.
10. H. S. Wang and N. Moayeri, Finite-state Markov channel—A useful model for radio communication channels, *IEEE Transactions on Vehicular Technology*, vol. 44, no. 1, pp. 163–171, 1995.
11. P.C. Reiss, M.W. White, Household Electricity Demand, Revisited (December 2001). NBER Working Paper No. w8687. Available at SSRN: http://ssrn.com/abstract=294736
12. A.-H. Mohsenian-Rad, V. W. S. Wong, J. Jatskevich, and R. Schober, Optimal and autonomous incentive-based energy consumption scheduling algorithm for smart grid, in *Innovative Smart Grid Technologies (ISGT), 2010*, January 2010, Gaithersburg, MD, pp. 1–6.
13. Office of Electricity Delivery and Energy Reliability, "Benefits of Demand Response in Electricity Markets and Recommendations for Achieving Them," U.S Department of Energy, 2006.

4

ADVANCED METERING INFRASTRUCTURE AND ITS INTEGRATION WITH THE DISTRIBUTION MANAGEMENT SYSTEM

ZHAO LI, FANG YANG, ZHENYUAN WANG, AND YANZHU YE

Contents

Recognizing the value of an advanced metering infrastructure (AMI), utilities worldwide are deploying millions of smart meters. To better justify AMI investment, researchers have recognized the urgency of utilizing the full potential of AMI to improve the quality of distribution management system (DMS) applications. However, the integration of AMI and DMS is a challenge as it entails different communication protocols and requirements for handling various meter information models. In addition, the AMI meter data load generated by millions of smart meters can potentially overwhelm DMSs. In this chapter, we first briefly review the state of the art of AMI technologies and then propose a unified AMI and DMS integration solution that easily adapts DMS systems to various AMI systems with minimal engineering effort.

4.1 Introduction

The advanced metering infrastructure (AMI)[1] consists of metering, communication, and data management functionalities, offering the two-way transportation of customer energy usage data and meter control signals between customers and utility control centers. AMI was originally developed from advanced meter reading (AMR),[2-6] a one-way communication infrastructure that implements automatic collection of meter measurements from residential smart meters to utility control centers for calculating monthly bills and fulfilling other related activities. Partially as the next generation of "AMR," AMI not only enhances the traditional data collection functionality (i.e., improving monthly meter data collection to real-time or near-real-time meter data collection) but also develops the communication capability from the control center to smart meters.

A distribution management system (DMS) is defined as an online decision-making tool that receives information pertaining to the system status and analog points from the distribution grid and generates supervisory control commands that are relayed to distribution breakers, switches and reclosers, switched capacitor banks, voltage regulators, and load tap changers (LTCs). To fulfill these functionalities, the DMS must have an efficient communication system capable of gathering the system state information and distributing control commands to customer-side control units (i.e., switches and reclosers) in real time and near real time.[7] Practically, however, because such a transportation network does not yet exist, most DMS applications (i.e., balanced or unbalanced load flow) are currently based on estimation values of data points, which leads to imprecise, even inaccurate, results.

In the past few years, AMI technologies have benefited from the U.S. government's economic stimulus plan. In addition, the Energy Policy Act of 2005 requires electric utilities with annual sales greater than 500,000 MWh to adopt the smart metering option with time-based rates. Today, most U.S. states have begun the process of deploying smart meters within an AMI. At the beginning of 2009, for example, Texas initiated a project of deploying 6 million smart meters and expected to complete it by 2012; California plans to install 10 million smart meters by the end of 2012. The deployment of smart meters is taking place not only in the United States but also throughout the world. Based on current estimates, by 2015 smart meter installations are expected to reach 250 million worldwide.[8] Hence, AMI and smart meters should be ubiquitous everywhere in the near future.

The deployment of AMI technologies has led to a need for a higher-quality DMS. Thus, the goal of research must be to integrate AMI with DMS systems.[9–12] As the intention of AMI was to serve a general domain that included electricity, water, and gas utilities, while that of the DMS was to exclusively serve the electricity domain, the integration of the two systems certainly entails the adaption of various communication protocols (i.e., American National Standards Institute [ANSI] C12.22, JMS [Java Messaging Service], and Web Service) and information models (i.e., ANSI C12.19, International Electrotechnical Commission [IEC] 61968-9, and MultiSpeak®) to the AMI and DMS systems. With the "tsunami" of AMI meter data generated by millions of residential smart meters, the task of integration has become even more complicated, requiring that the integration solution be scalable enough to handle the influx of a large number of meter measurements.

The rest of this chapter is structured as follows: The second section analyzes the components of the AMI (smart meters, the communication network, and the meter data management system [MDMS]) and reviews the current status and future trends of these components. The third section discusses the standardization of the AMI meter data model and communication protocols, an effective way to protect a utility's long-term investment in the AMI by extending the life cycle of AMI. The fourth section discusses the challenges in the integration of DMS and AMI integration; based on the discussion in this section, the fifth section conducts a meter data integration (MDI) case study. The last section concludes the chapter.

4.2 The Advanced Metering Infrastructure

The AMI consists of a metering system, a communication network, and an MDMS. In this chapter, we briefly review the functionalities and future trends of these AMI components.

4.2.1 The AMI Metering System

As the end device of the AMI, the AMI metering system refers to all electricity meters, which perform both measuring and communication functions, installed at customer sites. AMI metering systems fall into two categories: electromechanical meters and digital solid-state electricity meters.

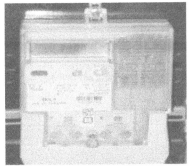

(a) Electromechanical meter   (b) Solid-state meter

Figure 4.1   (a) Electromechanical meter and (b) solid-state meter. (From Electromechanical meter and solid-state meter. http://en.wikipedia.org/wiki/Electric_energy_meter.[13])

An electromechanical meter (Figure 4.1) operates by counting the revolutions of an aluminum disk, designed to rotate at a speed proportional to the power. The number of revolutions proportional to energy usage determines the amount of energy consumption during a certain period. Currently, most utilities have a large number of electromechanical meters in the field that provide reliable and dependable measurement services. However, the major constraint of the electromechanical meter is its limited and nonexpendable measurement capabilities, which prevent its wide application in modern "smart" power grids.

A solid-state electricity meter, a meter constructed by digital signal-processing technologies, is actually a computer system that utilizes the microprocessor to convert analog signals to digital signals and further processes these digital signals into user-friendly results. For solid-state meters, adding a new function is as easy as installing a new application in a general computer. Hence, its functionalities can be easily expanded to adapt to various application scenarios. For example, beyond the traditional kilowatt-hour consumption measurement, a solid-state meter provides demand interval information, time-of-use (TOU) information, load profile recording, voltage monitoring, reverse flow and tamper detection, power outage notification, a service control switch, and other applications.

To communicate with other smart meters or utility control centers, a smart meter is generally equipped with a communication module. Popular communication modules in the current market are low-power

Table 4.1   Communication Modules of Primary AMI Vendors in the United States (March 2011)

AMI VENDORS	COMMUNICATION MODULES
Landis + Gyr	Unlicensed RF, PLC
Itron	ZigBee, unlicensed RF, public carrier network (OpenWay®)
Elster	Unlicensed RF, public carrier network
Echelon	PLC, RF, Ethernet
GE	PLC, public carrier network, RF
Sensus	Licensed RF (FlexNet®)
Eka	Unlicensed RF (EkaNet®)
Smart Synch	Public carrier network
Tantalus	RF (TUNet®)
Trilliant	ZigBee, public wireless network

Note: RF = radio frequency.

radios, the Global System for Mobile Communications (GSM), general packet radio services (GPRS), Bluetooth, and others. In general, each AMI vendor develops its own proprietary communication modules (Table 4.1) that are not interoperable with the communication modules produced by other vendors in most cases.

For most utilities, the deployment of millions of smart meters is a huge investment, so many utilities still maintain numerous electromechanical meters. However, because of their limited and non-expendable functionalities, the meters are gradually becoming a major obstacle to the utilities shifting to the smart grid, which requires a change in the functionalities of the end devices. Because of technological enhancements of the smart grid, utilities are gradually replacing their electromechanical meters, which they expect to last well into the future with solid-state meters, so the solid-state meters should begin to dominate the market in the near future.

4.2.2 AMI Communication Network

The AMI communication network is a two-way data transportation channel that transports meter measurements and meter control signals back and forth between individual meters and utility control centers. Technically, the AMI network can be categorized into either a hierarchical AMI network or a mesh AMI network. Because the mesh AMI network, a relatively new network, has several advantages

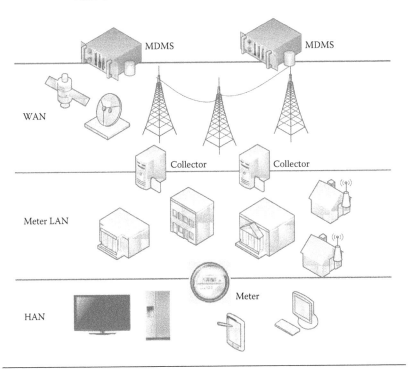

Figure 4.2 Infrastructure of the hierarchical AMI network.

(i.e., performance and efficiency) over the hierarchical AMI network, it will be the dominant AMI network in the future. Both types are discussed here.

4.2.2.1 The Hierarchical AMI Communication Network Format In a hierarchical AMI communication network, lower-level elements have strict relationships with their direct upper-level parent elements, and a meter is managed by its upper-level data collector. Figure 4.2 illustrates a typical multilevel hierarchical AMI communication network, which can be classified into three layers: the home-area network (HAN), the meter local-area network (LAN), and the wide-area network (WAN).[14,15] YY In such networks, meter data are collected and transported from a lower- to higher-level meter data collector. The major features of each layer are described next:

4.2.2.1.1 Wide-Area Network As the highest level of aggregation in an AMI network, the WAN handles connectivity between a high-level meter data collector and a utility control center or between high-level meter data collectors. The WAN is the backbone of the AMI communication network through which numerous AMI measurements and control signals are transported.

4.2.2.1.2. Meter LAN The meter LAN distribution network handles connectivity from data concentrators or some distribution automation devices (e.g., monitors, reclosers, switches, capacitor controllers) to high-level data collectors. Compared with the WAN, the meter LAN has larger geographical coverage but less data transportation.

4.2.2.1.3 Home-Area Network For utilities, the HAN has been defined or viewed as a grouping of home appliances and consumer electronic devices that allow for remote interface, analysis, control, and maintenance. The electric meter acts as the gateway of the HAN: collecting measurements (e.g., electricity, water, and gas) and sending them to the utility control center while executing control commands received from the utility control center.

The WAN, the LAN, and the HAN are generally constructed by wired and wireless network technologies. In the current AMI communication network, while widely applied wired communication technologies include communication via telephone systems, Ethernet, power line carriers, and broadband over power lines, widely applied wireless technologies include communication via mobile systems, cellular networks, and wireless mesh networks. Table 4.2 demonstrates the features of these network technologies in the current market.

The various layers of the AMI network require different performance, coverage, and security, so they are constructed by different

Table 4.2 Features of Primary AMI Communication Technologies

	NAME	DATA RATE	RANGE	SECURITY
Wired	PLC	100K bps	Same with power network	Exposed to public access
	BPL	<200 Mbps	Same with power network	Exposed to public access
	Fiber optic	10–40 Gbps	30–50 miles with repeaters	With security features
Wireless	WiMAX	<70 Mbps	Up to 30 miles	With security features
	Wi-Fi	11–54 Mbps	<100 m	With security features
	ZigBee (802.15.4)	20–250 kbps	<1 mile	With security features

wired or wireless communication technologies. For the HAN, which requires self-healing, plug-in play, low power, and low cost, ZigBee is the preferred technology. For the LAN, which requires good coverage and relatively low performance, power line communicator (PLC), unlicensed spectrum radio, and Wi-Fi are likely choices. For the WAN, which requires both high performance and good coverage, broadband over power line (BPL), WiMAX, and the licensed/unlicensed spectrum radio are preferable.

4.2.2.2 Internet-Protocol-Based Mesh AMI Communication Network   In a mesh AMI communication network based on the Internet Protocol (IP), a meter is an IP-based device capable of gaining access to meter data collectors and other meters through its IP address. In this sense, the IP-based mesh AMI network is similar to the Internet/intranet. Because of this similarity, many standard internet communication protocols (HTTP [Hypertext Transfer Protocol] and XML [eXtensible Markup Language]) are widely used in the IP-based AMI network even though they are neither specially designed nor optimized for utility meter data transportation. For example, WebGate Classic Residential Meter Solutions provided by MuNet[15] offers an IP-enabled AMI mesh network solution in which a meter can communicate with another meter or a meter data collector through standard Internet communication protocols (e.g., HTML and XML).

Compared with the multilevel hierarchical AMI network, a mesh AMI network has more advantages, especially in scalability, stability, and extensibility. More important, many well-developed and fully tested software and hardware technologies from the Internet (e.g., various communication protocols and network security technologies) can be smoothly transplanted into the AMI mesh network, making it more secure and user friendly. For example, Internet addressing technologies (e.g., IP version 4 [IPv4] and IPv6) help utilities effectively identify and control individual meters located in the network.

Generally, most advanced smart grid applications tend to transport a large amount of meter data in an efficient and secure way. Because existing hierarchical communications networks are incapable of performing such a task, the development of a more advanced AMI network is becoming urgent. As meshed AMI network technologies are still in the research-and-development (R&D) phase, an intermittent solution is to borrow matured Internet communication technologies and apply them to the existing AMI network. However, this solution is not tailored to power grid applications, so it must eventually be replaced by AMI mesh network technologies.

4.2.3 The Meter Data Management System

While a utility can use the AMI to collect data, it must also be able to use its AMI data to support decision making throughout the organization to achieve the maximum return on its investment. With the development of the smart grid, utilities are gradually realizing that the AMI cannot achieve all of the desired benefits unless it can effectively cleanse, process, store, and apply the data, activities that must be performed if utilities are to address and enhance their key business processes. These goals have driven the need for an entirely new MDMS.

The MDMS of the AMI provides a set of advanced software tools that manage large volumes of meter data. It collects, validates, and stores meter data in a central data repository and allows utilities to take full advantage of AMI information in: network monitoring, load research, operational analyses, and decision making. In addition, it enables meter data to be shared with end customers, who can access the data whenever they need to make decisions about how and when they use energy.

The typical functionalities of an MDMS are as follows:[16]

- Setting up, configuring, and monitoring meters and communication networks
- Administrating network security and data access privilege
- Loading meter data from communication gateways
- Providing a graphic user interface
- Cleaning, parsing, and storing data as well as exporting data to other systems
- Processing validated meter data for various utility applications

Since an MDMS collects meter readings from millions of meters at a certain time interval (i.e., 15 minutes), the volume of meter data is always increasing and potentially can become huge. Therefore, the challenge is to store and manage such a huge dataset and then extract valuable information from it to support various utility applications, tasks that cannot be controlled using traditional database technologies.

However, a well-defined solution that provides sufficient scalability to manage such a meter dataset is in the development phase in both theory and practice.

4.3 The Standardization of the AMI

In the current market, smart meters from different vendors are generally noninteroperable. For most utilities, deploying millions of smart meters is a long-term investment, which means that once a utility adopts smart meters from an AMI vendor, it must follow up with related products from the same vendor for the sake of compatibility. However, utilities are reluctant to be bound to a certain meter vendor, especially in the early stages of smart grid development.

Enabling interoperability between AMI products from different vendors is an effective way to protect utilities' investment, so most important standard committees in the world (e.g., AEIC [American Energy Innovation Council], ANSI, EPRI [Electric Power Research Institute], and NIST [National Institute of Standards and Technology]) are currently responding to this issue. Table 4.3 lists the popular standard communication protocols and meter information models in the current market, defined by ANSI, IEC, and NRECA (National Rural Electric Cooperative Association). Most of these standards have recently been revised (i.e., version 2 of C12.19 in 2008) or newly defined (i.e., version 1 of C12.22 in 2008) to support new requirements (i.e., demand responses) from the smart grid.

The standardization of the AMI includes standardization of both AMI communication protocols and AMI information models.

4.3.1 Standard AMI Communication Protocols

Since 2008, the focus of the standardization of AMI communication protocols has gradually shifted from the physical level (e.g., ANSI C12.18[17]) and the device level (e.g., ANSI C12.21[18]) to the application level (e.g., ANSI C12.22[19]) because the application-level communication protocols effectively isolate the details of underlying physical network configurations and implementations.

In the following section, we introduce the principal application-level communication protocols and meter information models that are

Table 4.3 Popular Standard Communication Protocols and Meter Information Models

	NAME	TIME TO MARKET	CATEGORY	FUNCTIONALITIES	APPLICATION DOMAIN
ANSI	C12.19	1997 version 1 2008 version 2	Data model	Model the meter data in tables	Gas, water, and electricity
	C12.22	2008 version 1	Communication protocol	Transfer data over C12.22 network	Gas, water, and electricity
	C12.18	1996 version 1 2005 version 2	Communication protocol	Transfer data by point-to-point protocol	Gas, water, and electricity
	C12.21	1999 version 1 2005 version 2	Communication protocol	Transfer data through a modem-based point-to-point protocol	Gas, water, and electricity
IEC	61968-9	2009 version 1	Data model	Model meter data for power system distribution application	Electricity
	62056	2007 version 1	Communication protocol	Transfer meter data over series port or network	Gas, water, and electricity
NRECA	MultiSpeak	Latest version 2007	Data model	Model meter data for power system distribution application	Electricity

popular in both the U.S. market (i.e., C12.19 and C12.22) and the European market (i.e., IEC 62056-53[20] and IEC 62056-62[21]).

4.3.1.1 ANSI C12.22 Historically, after a set of standard table contents and formats were defined in ANSI C12.19 (the details for the C12.19 standard are discussed further in this chapter), a point-to-point standard protocol (ANSI C12.18) was developed to transport the table data over an optical connection. The *Protocol Specification for Telephone Modem Communication* (ANSI C12.21) was developed afterward to allow devices to transport tables over telephone modems. The C12.22 standard expands on the concepts of both ANSI C12.18

and C12.21 standards to allow the transport of table data over any reliable networking communications system.

4.3.1.1.1 Goals of ANSI C12.22   The goal of the ANSI C12.22 standard is to define a meshed network infrastructure that is customized for AMI applications. The goals of the standard are as follows:

- To define a datagram that may convey ANSI C12.19 data tables through any network, which must include the AMI network and optionally includes the Internet
- To provide a seven-layer communication infrastructure for interfacing a C12.22 device to a C12.22 communication module
- To provide an infrastructure for point-to-point communication to be used over local ports such as optical ports or modems
- To provide an infrastructure for efficient one-way messaging

Overall, the ANSI C12.22 mesh network consists of the C12.22 nodes and network.

4.3.1.1.2 Network Infrastructure of ANSI C12.22   A C12.22 node, a point on the network that attaches to a ANSI C12.22 network (Figure 4.3), is a combination of both a C12.22 device and communication module. The C12.22 communication module is a hardware module that attaches a C12.22 device to a C12.22 network. The C12.22 device contains meter data in the forms of tables defined

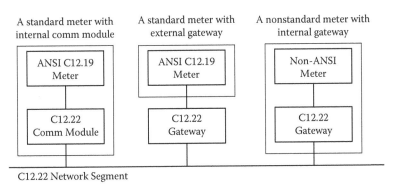

Figure 4.3   Typical examples of C12.22 nodes.

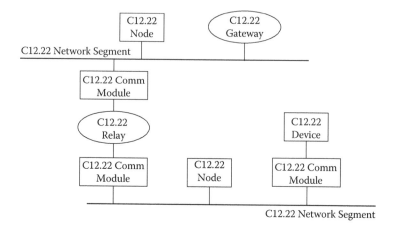

Figure 4.4 The basic C12.22 network.

in the C12.19. The interface between the communication module and the device is completely defined by the C12.22 standard.

The C12.22 network defines an AMI-specific mesh communication infrastructure that consists of one or more C12.22 network segments (a subnetwork) or a C12.22 LAN (Figure 4.4). Within a network segment, there is a collection of C12.22 nodes that communicate with one another without forwarding messages through either a C12.22 relay or a C12.22 gateway. The C12.22 network segments can be connected into a C12.22 WAN through C12.22 relays and gateways, where meters from different network segments can communicate with one another.

Similar to the Open System Interconnection (OSI) model, the C12.22 communication protocol consists of seven layers (Figure 4.5): an application layer (layer 7), a presentation layer (layer 6), a session layer (layer 5), a transport layer (layer 4), a network layer (layer 3), a data link layer (layer 2), and a physical layer (layer 1). Unlike OSI, C12.22 is customized for meter data transportation. For example, the application layer of C12.22 supports only ANSI C12.19 tables, EPSEM, and ACSE (EPSEM and ACSE are languages that encapsulate C12.19 meter data[22]). The standard services provided by layer 7 of C12.22 include an identification service, a read service, a write service, a security service, a trace service, and others; layers 1 through 6 support various physical network connections in the meter industry as well as the standard Internet connection.

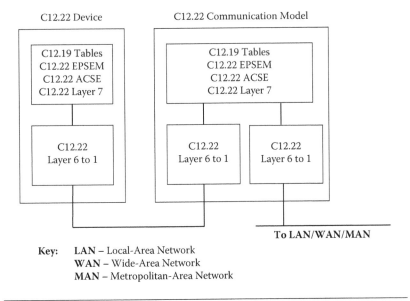

Figure 4.5 Seven-layer Open System Interconnection model for meter data transportation.

4.3.1.2 IEC 62056 IEC 62056, which defines the meter interface classes for the Companion Specification for the Energy Metering (COSEM) model, includes a series of standards on data exchange for meter reading, tariffs, and load control, as follows:

- IEC 62056-21: A standard that defines direct local data exchange, which describes how to use COSEM over a local port (optical or current loop). It is designed to operate over all media, including the Internet, through which a meter sends ASCII or other format meter data to a nearby handheld unit using a serial port.
- IEC 62056-42: A standard that defines physical-layer services and procedures for connection-oriented asynchronous data exchange.
- IEC 62056-46: A standard that defines a data link layer using the High-Level Data Link Control (HDLC) protocol, a three-layer, connection-oriented, HDLC-based communication profile.
- IEC 62056-47: A standard that defines COSEM transport layers for IPv4 networks, the Transmission Control Protocol [TCP]/IP-based communication profile.

- IEC 62056-53: A standard that defines a COSEM application layer.
- IEC 62056-61: A standard that defines an object identification system (OBIS).
- IEC 62056-62: A standard that defines interface classes and a data model.

Similar to ANSI C12.22, IEC62056-53, the application-layer communication protocol in the COSEM model (Figure 4.6), is defined based on several other IEC 62056 series protocols, including IEC 62056-21, −42, −46, and −47. Except for IEC 62056-21, which is used in handheld devices for locally exchanging data with meters, the remaining protocols are used to define different layers of the communication network that support application-level communication: the physical layer (IEC 62056-42), the data link layer (IEC 62056-46), and the transport layer (IEC 62056-47). Similar to ANSI C12.22, the meter data carried by IEC 62056-53 are defined by IEC 62056-61 and IEC 62056-62, which are dedicated meter data models in the IEC 62056 series.

As an application-layer communication protocol, IEC 62056-53 primarily provides three services to application-level semantics: the GET service (.request, .confirm), the SET service (.request, .confirm), and the ACTION service (.request, .confirm).

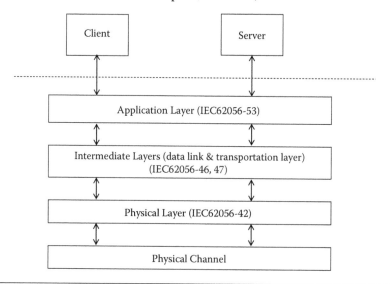

Figure 4.6 Request/response process of COSEM.

Although both IEC 62056 and ANSI C12.22 provide a way of constructing the advanced mesh AMI network, each has a unique market focus: IEC 62056 primarily focuses on the European market, while ANSI C12.22 focuses on the North American market. In the current North American market, most AMI vendors support C12.18 and C12.21, but few support C12.22 since it has only recently been defined. Itron,[23] Elstor,[24] and Trillant Incorporated[25] were the pioneers supporting the C12 communication protocols. Because of the advantages of C12.22, we predict that in the near future most major meter vendors will support C12.22 standard communication protocols in the North American market.

4.3.2 Standard AMI Information Model

An information model is a representation of concepts, relationships, constraints, rules, and operations that specify data semantics for a chosen domain of discourse.[26] In the AMI communication infrastructure, it is necessary that an information model, in which all communication participants can semantically reach a certain level of understanding, be maintained.

In this chapter, we briefly discuss major standard information models in today's market: ANSI C12.19 and IEC 62056-62. The former is widely used in the U.S. market and the latter in the European market.

4.3.2.1 ANSI C12.19-2008   ANSI C12.19 resulted from comprehensive cooperative effort among utilities, meter manufacturers, automated meter-reading service companies, ANSI, Measurement Canada (for Industry Canada), NEMA, the IEEE (Institute of Electrical and Electronics Engineers), Utilimetrics, and other interested parties. Currently, it has two versions: ANSI C12.19-1997 and ANSI C12.19-2008. As the latter is intended to accommodate the concepts of the most recently identified AMI, it is primarily discussed in this chapter.

The heart of ANSI C12.19 is a set of defined standard tables and procedures; the former are methods of storing the collected meter data and controlling parameters, and the latter are methods of invoking certain actions against the data and parameters.[22] The standard tables in C12.19 are typically classified into sections, referred to as decades. Each decade pertains to a particular feature set and a related

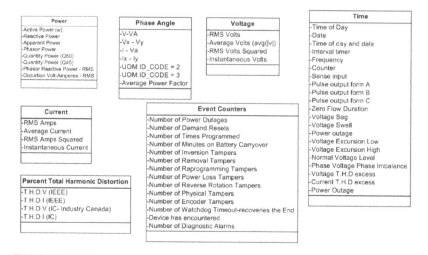

Figure 4.7   Electricity information modeled by C12.19. THD = total harmonic distortion, V = voltage, I stands for current.

function. Transferring data from or to an end device that adheres to the C12.19 standard entails reading or writing a particular table or a portion of a table. Even though the C12.19 standard covers a broader range of tables and procedures, it is highly unlikely that any smart meter will be able to embed all tables or even a majority of those defined in ANSI C12.19. Hence, implementers are encouraged to choose an appropriate subset that suits their needs.

C12.19 is a general meter information model that serves various domains, including electricity, water, and gas. As an example, Figure 4.7 illustrates the electricity information abstracted from the tables defined in Decade 1 of the C12.19 standard. In addition, the tables in C12.19 can be customized through some standard operations.

4.3.2.2 IEC 62056-62   Unlike ANSI C12-19, which uses tables to package meter measurements, IEC 62056-62 models meter information through a series of interface classes.[21] As the information modeled by C12.19 and IEC 62056-62 is identical, we do not duplicate our efforts to further introduce the content of IEC 62056-62. Similar to ANSI C12.19, as a general meter data model, IEC 62056 supports not only electricity meters but also gas and water meters.

For AMI vendors, the preference to support certain standards has a strong geographical bias. For example, most smart meter vendors

Table 4.4   Relationships between IEC 62056 Series Standards and Primary Meter Vendors in the U.S. Market (March 2011)

	IEC 62056/DLMS/COSEM
Landis + Gyr	Europe: IEC 62056-21 (for local reading) and DLMS (as a system integration interface)
North America: No	
Itron	United States: Quantum: mini-DLMS
	Europe: IEC 62056-21 and DLMS/COSEM for C&I meter
Elster	A1800 ALPHA: DLMS/COSEM and IEC 62056-42, −46, −53, −61, −62
Echelon	IEC 62056-21 (2002) (physical and electrical requirements only)
GE	No
Sensus	No
Eka	No
SmartSynch	No
Tantalus	No
Trilliant	No

Source: International Electrotechnical Commission. With permission.

in the U.S. market are more likely to choose ANSI series standards (i.e., C12.19 and C12.22), while those in the European market are more likely to select IEC standards. Table 4.4 lists the situations of the major meter vendors in the U.S. market that support the IEC 62056 series standards. As of today, only Elster completely supports IEC 62056 series standards, including IEC 62056-42, −46, −53, −61, and −62. Other vendors, such as Itron, support only a portion of the IEC 62056 standards, and some such as GE and Sensus do not support the series protocols at all.

Triggered by the rapid development of the smart grid, beyond supporting proprietary communication protocols, most AMI vendors have begun to support the standard communication protocols and meter data models. As of today, most vendors have accepted C12.19 (Table 4.5), but only a few pioneers (i.e., Itron and Elster) support C12.22, which is necessary for a future meshed AMI network.

Overall, AMI is a two-way communication network ranging from residential houses to control centers. As an information provider, it is complementary, to some extent, to DMS, providing real-time or near-real-time system state information, and as a command executor, conducting control commands sent from the utility control centers to residential smart meters. As real-time or near-real-time system state

Table 4.5 AMI Vendors and Standard Information Models and Communication Protocols in the U.S. Market (March 2011)

	C12.18	C12.21	C12.22	C12.19	IEC 61968/ CIM	OTHERS
Landis + Gyr	V	V	V	V	V	Unlicensed RF, PLC
Itron	V	V	V	V	V	ZigBee, unlicensed RF, public carrier network (OpenWay®)
Elster	V	V	V	V	V	Unlicensed RF, public carrier network
Echelon	V			V	V	PLC
GE	V	V		V	V	PLC, public carrier network, RF
Sensus	V	V		V		Licensed RF (FlexNet®)
Eka	V			V		Unlicensed RF (EkaNet®)
SmartSynch			V	V		Public carrier network
Tantalus	V	V	N/A	V	N/A	RF (TUNet®)
Trilliant	V	V	V	V	Not yet	IEEE 802.15.4; ZigBee; public WAN, including CDMA/1xRTT, GSM/GPRS, WiMAX, etc.

Note: CDMA = code division multiple access; DLMS = Distribution Line Message Specification.

information can significantly improve the quality of DMS applications, integration of the AMI with the DMS may represent a feasible, efficient solution for improving the quality of DMS applications.

4.4 The AMI and DMS Integration

In this section, we focus on the context, issues, and challenges of AMI and DMS integration from an engineering aspect.

4.4.1 Meter Data Models in the DMS

Instead of adopting existing AMI meter data models (i.e., C12.19 and IEC 62056-62), the DMS defines its own meter data models that are exclusively optimized for DMS applications and are compatible with existing DMS information models (e.g., the Common Information Model [CIM]). The most popular meter data models in the DMS today are IEC 61968-9 and MultiSpeak.

4.4.1.1 IEC 61968-9: A Meter Model in CIM   Published by IEC in 2009, the IEC 61968-9[27] standard defines the interface for meter reading and control in the DMS. The goal of the interface is the exchange of information between a meter system and other applications at electric utilities, serving the integration of meter data with utility applications. As part of the CIM of the utilities, the IEC 61968-9 standard extends the traditional CIM to support the exchange of meter information between utility applications. Electricity measurements provided by IEC 61968-9 are important for a variety of DMS applications (Figure 4.8), such as outage management, service interruptions, service restoration, quality-of-service monitoring, distribution network analysis, distribution planning demand reduction, customer billing, and work management.

In addition to electricity measurements, the IEC 61968-9 standard defines a meter information exchange infrastructure consisting of message and event definitions, which are meter reading, meter control, meter events, customer data synchronization, and customer switching.

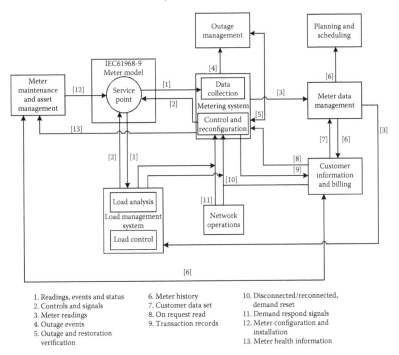

1. Readings, events and status
2. Controls and signals
3. Meter readings
4. Outage events
5. Outage and restoration verification
6. Meter history
7. Customer data set
8. On request read
9. Transaction records
10. Disconnected/reconnected, demand reset
11. Demand respond signals
12. Meter configuration and installation
13. Meter health information

Figure 4.8   Improving the quality of DMS applications using AMI meter data.

```xml
<?xml version="1.0" encoding="UTF-8"?>
<!--edited with XMLSPY v2004 rel. 3 U (http://www.xmlspy.com) by ABB
 (ABB Inc) -->
<m:MeterReadings
 xsi:schemaLocation="http://iec.ch/TC57/2009/MeterReadings#
  MeterReadings.xsd"
xmlns:m="http://iec.ch/TC57/2007/MeterReadings#"
xmlns:xsi="http://www.w3.org/2001/XMLSchema-instance">
            <m:MeterReading>
                      <m:MeterAsset>
                                  <m:mRID>6468822</m:m|RID>
                      <m:MeterAsset>
            <m:MeterReading>
</m:MeterReadings>
```

Figure 4.9 Example of the meter pull information packaged by IEC 61968-9. (With permission from the IEC.)

Figure 4.9 demonstrates an IEC 61968-9 message sent by DMS polling a meter reading based on a meter ID from an AMI system.

As part of a CIM, a meter modeled by IEC 61968-9 is represented by the MeterAsset class, a newly defined class in the CIM that supports smart meters. Through the MeterAsset class, an IEC 61968-9 meter can easily exchange information with other devices modeled by the CIM and provide better services for utility applications. More important, unlike C12.19, which is a general meter data model serving water, gas, and electricity, a meter in IEC 61968-9 is exclusively tailored to utility applications (e.g., load analysis and control, outage management and meter maintenance, and asset management).

4.4.1.2 MultiSpeak MultiSpeak[28] is a de facto standard funded by NRECA. Similar to IEC 61968, it focuses on data exchange modeling and enterprise integration in electric utilities and is intended to support standards-based interapplication integration. Compared to IEC 61968-9, MultiSpeak is a mature protocol that has been in the market for some time.

From an infrastructure perspective, IEC 61968-9 fits into a variety of messaging middleware frameworks, so it is suitable for utilities that may have a number of different middleware solutions already in place. MultiSpeak, implemented in terms of Web services, is more effective for small utilities, which rarely implement messaging middleware (Figure 4.10).

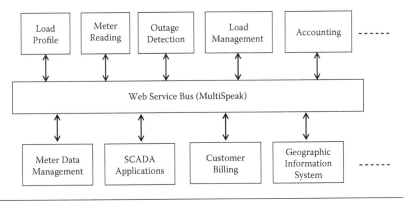

Figure 4.10   MultiSpeak data model.

4.4.1.3 Comparison of General Meter Models and Power System, Domain-Specific Meter Models   Overall, ANSI C12.19 and IEC 62056-62 are general meter models serving the electricity domain, the water domain, and the gas domain, while MultiSpeak and IEC 61968-9 are customized to the electricity domain, which supports enterprise integration within the scope of the electric utilities.

Figure 4.11 illustrates the relationships between general meter models and power system-specific meter models. Even though all

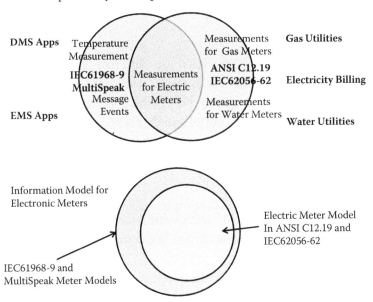

Figure 4.11   General meter data model versus power system-specific data model.

four meter data models discussed have electricity measurements, the scopes of the measurement namespaces and message infrastructures differ (refer to the lower part of Figure 4.11). As electricity-specific models (i.e., IEC 61968-9 and MultiSpeak) organize and model information based on utility applications, they have a larger scope of meter measurement namespaces that include temperature measurements, demand response control, and other domain information, providing better services for various DMS applications. In addition, the message and event infrastructure of electricity meter models are compatible with the existing utility information infrastructure, so they can easily be integrated with utility applications.

4.4.2 AMI and DMS Integration

4.4.2.1 Business Considerations   The DMS and the AMI are two separate systems with different business goals and architectures. The purpose of integrating the AMI and the DMS is to enable the two systems to exchange information while minimizing the influence of the integration on both systems in terms of performance and engineering costs.

Utilities that adopt the MDI layer must address some major business considerations:

- Different DMS applications use different approaches to import external data. For example, some DMS applications utilize the enterprise service bus (ESB), and others rely on the SCADA (supervisory control and data acquisition) system. In this sense, the integration solution should be easily adaptable to both ESB and SCADA interfaces.
- Utilities generally have deployed (or are currently developing) AMI systems with different types of meter data servers such as MDMSs. These AMI systems are usually built by different AMI vendors. The integration solution should be adaptable to diverse AMI systems.
- Meter data models in AMI systems are designed for applications in different domains (i.e., electricity, water, and gas); however, DMS applications primarily require a power system-dedicated meter data model (i.e., IEC 61968-9). The

integration solution should consider information gaps between a general domain meter data model and a power system—a specific meter data model—and match these gaps while an AMI system and a DMS system are exchanging meter data.

- Currently, the commercial requirement for regular meter reading is a 15-minute interval. With the development of AMI technologies, the interval is becoming shorter. However, handling meter data generated by millions of smart meters every 15 minutes can pose substantial challenges to DMS systems because the original architecture of most legacy DMS systems was not designed for heavy AMI meter data load conditions. Therefore, the integration solution has to minimize the influence of the meter data load on a DMS system by caching meter data and adjusting the meter data stream throughput to a level that can be accepted by the DMS system when necessary.

- AMI and DMS systems are usually developed by different vendors. Thus, a different dialect is likely used in both meter models and transportation protocols to describe even the same grid network. The integration solution should harmonize these dialects.

4.4.2.2 Challenges of AMI and DMS Integration The challenges of designing the integration solution are as follows:

- Performance: The integration solution should be capable of processing both scheduled (expected) meter reading data and burst occurred (unexpected) meter outage reports generated by a large number of smart meters in a timely manner. Ideally, it should be able to withstand a worst-case data-loading scenario when a regular meter reading session coincides with a large-scale outage reporting event.

- Scalability: The integration solution should be able to handle the incoming data of thousands and even millions of smart meters at a data-updating interval of 1 day, 1 hour, 15 minutes, or even shorter.

- Adaptability: The integration solution should be able to adapt to distinct AMI systems deployed by utilities. That is, it should be capable of adapting to different meter data models and different AMI data integration communication protocols.

• Extensibility: The integration solution should be capable of introducing appropriate technologies that fit not only the current AMI and DMS technology framework but also future AMI and DMS applications.

4.5 The Meter Data Integration Layer: A Unified Solution for the AMI and DMS Integration

4.5.1 The Context of the MDI Layer

A unified AMI and DMS integration solution, called the MDI layer,[29] is described in this section; it can be viewed as middleware between the AMI and DMS systems. Figure 4.12 illustrates the overall system, which consists of the AMI, the DMS, and the integration solution. The MDI layer enables the easy integration of diverse AMI systems (i.e., the MDMS meter data collection engine) with the DMS.

4.5.2 Software Architecture of the MDI Layer

In this section, the MDI layer architecture and the rationale and trade-offs behind the architecture are discussed.

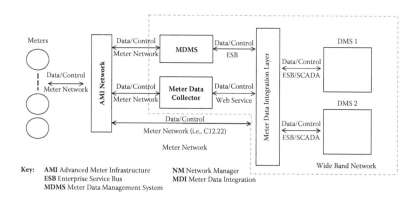

Figure 4.12   Context of the MDI layer. (From Z. Li, Z. Wang, et al., *2010 First IEEE Smart Grid Communication*, NIST, Washington, DC, October 2010, pp. 566–571. Copyright IEEE. With permission from IEEE.)

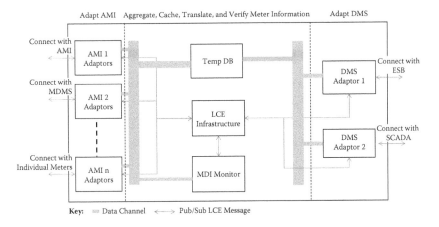

Adapt AMI   Aggregate, Cache, Translate, and Verify Meter Information   Adapt DMS

Figure 4.13 Architecture of the meter data integration solution. (From Z. Li, Z. Wang, et al., *2010 First IEEE Smart Grid Communication*, NIST, Washington, DC, October 2010, pp. 566–571. Copyright IEEE. With permission from IEEE.)

4.5.2.1 Components of the MDI Layer Figure 4.13 is the component diagram of the MDI layer. Overall, the components of the MDI can be classified into four categories based on their functionalities: the AMI adaptors, the AMI information translation and verification infrastructure, the loosely coupled event (LCE) infrastructure, and the DMS adaptors.

4.5.2.1.1 AMI Adaptors AMI adaptors, the components illustrated on the left side of Figure 4.13, can fit different types of AMI data servers (e.g., the MDMS and the meter data collector), transferring metering data streams from the AMI to the DMS or vice versa:

- For transferring meter data (e.g., measurements and outage information) from the AMI to the DMS, AMI adaptors must conform to the AMI communication channel to receive the meter data blocks sent by the corresponding AMI server and process them. To fulfill this conformable process, AMI adaptors must understand the communication protocols and meter data models of the AMI system involved.
- For transferring metering information (e.g., meter control and meter poll commands) from the DMS to the AMI, AMI adaptors must pack the meter poll/control information using the AMI meter data model and deliver the packages to the AMI system using the communication protocol adopted by the AMI system.

Each AMI system that needs to support DMS applications should have a corresponding AMI adaptor in the MDI layer. An ideal AMI adaptor possesses the following attributes:

- It has a high-performance parser that processes the incoming AMI meter data stream effectively.
- It can be dynamically plugged into the MDI layer without interrupting the normal operation of other components in the MDI layer.

The utilization of AMI adaptors can greatly simplify the process of adapting a DMS system to diverse AMI systems because, for a new AMI, the deployment process requires a simple redesign of a corresponding lightweight AMI adaptor instead of a redevelopment of the entire AMI interface.

4.5.2.1.2 The Information Translation and Verification Structure   Due to the variations among the meter data models with regard to typical AMI and DMS applications, we concluded in Wang and Li[30] that AMI systems utilize general meter data models (i.e., ANSI C12.19) that can be applied to all domains (i.e., electricity, gas, and water). However, DMS systems use power system-specific meter data models (i.e., IEC 61968-9), so one of the primary tasks of the MDI layer is to eliminate the information gaps between the AMI and DMS meter data models, which is accomplished using the information translation and verification structure, the components of which are shown in the middle part of Figure 4.13.

In the MDI layer, translation means converting the AMI dialect to the DMS dialect when meter data from the AMI are delivered to the DMS and vice versa. Translation is implemented by looking up the AMI and the DMS cross-reference tables stored in the TempDB. Verification involves filtering error information and verifying the integrity of the incoming metering data before delivering them to the target system (either the AMI or the DMS). Verification is implemented by the foreign key constraints of the relational tables in the TempDB.

More important, in the worst-case scenario, in which a regular meter-reading session coincides with a large-scale outage event, when a large amount of incoming meter information requires verification in a short time frame, the information translation and verification infrastructure guarantees performance by utilizing threading technology.

4.5.2.1.3 The LCE Infrastructure The LCE infrastructure is the messaging infrastructure of the MDI layer. All functional components in the MDI layer are coordinated by publishing or subscribing messages to the LCE infrastructure. The LCE event infrastructure[31] has two characteristics:

1. It is a messaging system that efficiently coordinates the behaviors of message senders/receivers.
2. It is loosely coupled. Message senders (publishers)/receivers (subscribers) of the LCE infrastructure are decoupled and run in different program spaces; shutting down a publisher or subscriber will not influence the normal operations of other publishers or subscribers.

4.5.2.1.4 DMS Adaptors Most design considerations for AMI adaptors (e.g., dynamically plugging in) are also applicable to the design of the DMS adaptors. More important, the design of the DMS adaptors also considers the throughput limitation of DMS data channels when delivering meter data to the DMS system.

4.5.2.1.5 MDI Monitor The MDI monitor, which is used to track the status of the functional components in the MDI layer by subscribing to the LCE messages sent by these components, can be dynamically plugged into the MDI by turning on its subscriptions to LCE messages.

4.5.2.2 Behavior of the MDI Layer Dynamically, the MDI layer supports the following three types of activities or events: (1) The AMI pushes meter data to the DMS (i.e., reports outage and pushes regular meter reading); (2) the DMS polls meter data from the AMI (i.e., verifies outage and requires meter measurements) by sending meter poll commands to the AMI; and (3) the DMS pushes control commands to the AMI (i.e., meter-controlling demand).

4.5.2.2.1 AMI Pushes Meter Data to the DMS   The workflow of the MDI layer processing meter data pushed by the AMI is as follows: On receiving a meter data package from the AMI, an AMI adaptor parses and delivers the parsed meter data to the TempDB for translation and verification; meanwhile, it publishes a message in the LCE infrastructure to notify other components in the MDI layer that are interested in the AMI meter data arrival notice. One of the other interested components is a DMS adaptor. On receiving this notice, the DMS adaptor initiates the following workflow to process the AMI meter data that have come in: First, it picks up the verified and translated meter data from the TempDB, then packs them using the message format required by the DMS system, and finally delivers the packed message to the DMS system. During the process, the behaviors of the AMI adaptor, the TempDB, and the DMS adaptor are coordinated by the AMI meter data arrival notice.

As previously mentioned, the meter data load pushed by the AMI can be very large. Thus, avoiding a delay in processing or a loss of AMI meter data requires a high-performance AMI adaptor. Performance can be enhanced by multicore and multithread technologies. The workflow of multithreaded AMI adaptors that process income meter data is as follows: On receiving a meter data package, the AMI adaptor quickly unpacks it and then concurrently launches a new thread that parses the meter data package, caches the meter data into the TempDB, and sends a meter data arrival notice to the LCE infrastructure. However, the creation of new threads consumes a considerable number of system resources. After all, when there is a large volume of data packages coming in, AMI adaptors must launch a large number of threads in a very short time frame, quickly using up system resources. This phenomenon is called a *thread explosion*. To prevent such an event from occurring, a semaphore installed in the AMI adaptor limits the number of launched threads.

4.5.2.2.2 DMS Pushes Meter Control Commands to the AMI   The workflow of processing meter control commands pushed by the DMS to the AMI is similar to that of processing meter measurements pushed by the AMI, except for their starting points: DMS adaptors for the former and AMI adaptors for the latter.

4.5.2.2.3 DMS Polls Meter Data from the AMI The workflow of the DMS polling meter data from the AMI consists of the following processes: (1) The DMS pushes meter control commands to the AMI, and (2) the AMI pushes the meter data back to the DMS. These two processes were discussed previously in this chapter.

4.5.3 *The MDI Architecture Evaluation*

To validate the MDI design and evaluate its quality attributes in a real-world situation, we developed a prototype of the MDI layer and a meter data load simulation system (the AMI simulator) using Microsoft.NET Enterprise technologies, a simulation system that can create various testing scenarios for the MDI layer by simulating many smart meter operations (i.e., meter outages, meter reading, and meter control). Using the MDI prototype and the AMI simulator, we ran several test cases to evaluate the functionalities and the associated quality attributes of the designed MDI layer.

4.5.3.1 Strategies Instead of exhausting all of the possible testing scenarios, we chose to test some important functionalities and their associated quality attributes. Because the most important attribute of the MDI layer is its ability to handle meter data loads pushed by millions of smart meters, we focused our test cases on a meter data "tsunami" scenario. In other words, we primarily tested the quality attributes (performance, scalability, and flexibility) of the architecture of the MDI layer against a meter data load created by millions of smart meters. The test environment is illustrated in Figure 4.14, and a detailed description of the quality attributes and their trade-offs follows.

4.5.3.1.1 Performance Typically, a utility has millions of smart meters whose data load can come from regular measurements every 15 minutes or outage reports caused by a burst. However, the available AMI simulator server (constrained by its central processing unit [CPU] and internal memory) can simulate the behavior of only 65,000 smart meters. Instead of simulating a meter data load generated by millions of smart meters in 15 minutes, we simulated the load of 63,445 smart meters (the meters corresponded to customers in a fixed

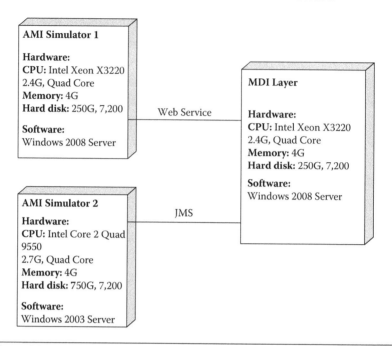

Figure 4.14 The configuration of the simulation test environment. (From Z. Li, Z. Wang, et al., A Unified Solution for Advanced Metering Infrastructure Integration with a Distribution Management System, *2010 First IEEE Smart Grid Communication,* NIST, Washington, DC, October 2010, pp. 566–571. Copyright IEEE. With permission from IEEE.)

number of areas in one utility network) in 1 minute. Our assumption was that if the MDI layer can process the meter data load generated by the 63,445 smart meters in 1 minute, then it should be able to process a meter data load generated by 1 million meters in 15 minutes.

4.5.3.1.2 Scalability We tested the scale-up and scale-out capabilities of the MDI layer. For the scale-up test, we used two AMI simulators that concurrently pushed outage reports to the same MDI layer server; accordingly, two AMI adaptors configured in the MDI layer server received the outage reports pushed by two AMI simulators. Unlike a typical AMI adaptor situation, the scale-up test case handled about double the meter data load.

For the scale-out test, we added a second MDI layer server, configured to use the same TempDB with the first MDI layer. In the two MDI layer server configurations, two AMI simulators were connected to the two MDI layer servers. The primary purpose for the scale-out test case was to verify if the designed MDI layer could be scaled out simply by modifying the server configuration rather than by revising the source code of the MDI layer.

4.5.3.1.3 Flexibility   For flexibility, we wanted to verify the capability of the MDI layer connected to different types of AMI systems. To make more sense of the test results, we simulated two AMI systems: one transported meter data using JMS,[32] supported by IBM WebSphere MQ 7.0 (transactional communication channel),[33] and the other AMI system transported meter data using Web service Remote Call (a nontransactional communication channel). Accordingly, the MDI layer had two adaptors, one for each AMI system.

4.5.3.2 Test Results and Discussion   The performance test results showed that the MDI layer server could cache, translate, and verify either the meter measurements or the outage information generated by 63,445 smart meters in 30 seconds. Based on this "half-minute" meter data load, we can calculate that one MDI layer server with similar resources can process the meter load generated by 1.9 million smart meters in 15 minutes.

The scalability test results showed that the designed MDI layer could easily be scaled up by adding a second AMI adaptor that is identical to the original AMI adaptor by changing the configurations rather than the source code or the design. The addition of the second AMI adaptor barely influenced the performance of the previous AMI adaptors. However, utilization of system resources increased; for example, CPU utilization increased from 50% to 70%. This demonstrated from another angle that if we have sufficient system resources (CPU and memory), we can connect two or more AMI systems within one physical MDI layer server. Hence, an optimistic estimation is that the scale-up of a configuration with two identical AMI adaptors can handle a meter data load generated by 3.8 million (2×1.9 million) smart meters in 15 minutes if the system resources of the test MDI layer server are adequate.

The flexibility test results showed that the designed architecture could easily connect to different AMI systems. Similar to the scale-up test, the flexibility test, which connected two different types of AMI systems (Web service and JMS), barely affected the performance of the MDI layer. In addition, adding a second different type of AMI adaptor increased the utilization of system resources, demonstrating that the MDI layer can be scaled up by connecting it to different AMI systems.

4.6 Conclusion

The revolutionary contribution of AMI is that it creates a low-cost standard communication network facilitating the collection and distribution of metering information to customers, utilities, and other parties. Because of AMI, a wide range of new DMS applications that used to be considered important but impractical due to communication costs (i.e., automatic outage management and demand responds) is introduced or reactivated. In return, these applications expose the precise state of the power distribution infrastructure and operational awareness for the optimization of the delivery and use of energy to utility control centers.

To effectively integrate AMI with DMS, a unified AMI and DMS integration solution, called the MDI layer, was presented in this chapter. Structurally a type of middleware deployed between the AMI and DMS systems, the MDI can greatly reduce development and engineering efforts expended connecting DMS applications to various types of AMI systems. At the same time, it can minimize the influence of the AMI meter data load on the performance of DMS applications by decoupling the data model and protocol conversion functionality from the DMS applications. More important, the MDI layer can be easily expanded by adding new functionalities (e.g., power system load profiling, forecasting and modeling, outage scooping, and asset utilization optimization) to fulfill requirements from potential DMS applications in the future.

As the quality attributes of the MDI layer in a real-world environment are a major concern, a series of test cases were conducted against an MDI prototype and an AMI simulator to verify the performance, flexibility, and scalability of the proposed MDI layer. The test results showed that the MDI layer design could meet the real-world requirements of handling AMI meter data generated by millions of smart meters in terms of performance, flexibility, and scalability.

In summary, with the development of the smart grid, AMI, as the backbone of information collection and distribution in the grid system, is gradually transcending the electric territory, expanding the network to millions of utility consumers, various renewable resources, and millions of electrical vehicles to the far edges of the delivery system, the initial prototype of the "Internet of things."

Acknowledgments

This work was supported by the ABB Corporation research funds that come from both the industry software system program and the grid automation program. In addition, we would like to thank Xiaoming Feng for valuable comments on the early version drafts.

References

1. F. Yang, *Advanced Metering Infrastructure Technology*, Prestudy Report No. PT-07045. Raleigh, NC: ABB U.S. Corporate Research Center, 2007.
2. R. A. Fischer, A. S. Laakonen, and N. N. Schulz, A generation polling algorithm using a wireless AMR system for restoration confirmation, *IEEE Transactions on Power Systems*, Vol. 16, No. 2, pp. 312–316, 2001.
3. H. Dorey, Advanced metering in old and new worlds, *Power Engineering Journal*, Vol. 10, No. 4, pp. 146–148, August 1996.
4. Y. Jin and M. D. Cox, A pipelined automatic meter reading scheme, paper presented at the Instrumentation and Measurement Technology Conference, Irvine, CA, pp. 715–720, May 1993.
5. S. Mak and D. Radford, Design considerations for implementation of large scale automatic meter reading systems, *IEEE Transactions on Power Delivery*, Vol. 10, No. 1, pp. 97–103, 1995.
6. M. R. J. Clay and A. J. McEntee, Advanced meter reading tokenless prepayment, *Power Engineering Journal*, Vol. 10, No. 4, pp. 149–153, August 1996.
7. Electric Power Research Institute. *The Introduction of Smart Grid*, 2007. http://www.epri.com/IntelliGrid/.
8. Pacific Gas and Electric (PG&E), *SmartMeter™ Installation Progress*, PG&E, April 2010. http://www.pge.com/myhome/customerservice/meter/smartmeter/deployment/.
9. R. W. Uluski, Interactions between AMI and DMS for efficiency/reliability improvement at a typical utility, paper presented at IEEE PES General Meeting, Raleigh, NC, July 2008.
10. Ali Ipakchi, Implementing the smart grid: Enterprise information integration, Grid-Interop Forum, 2007.

11. General Electric, *Advanced Distribution Infrastructure, GE's AMI and DMS Integration Solution.* http://www.gepower.com/prod_serv/products/metering/en/going_ami_new.htm

12. EnergyIP, *Siemens's AMI and DMS Integration Solution.* http://www.energy.siemens.com/us/pool/us/services/powertransmission-distribution/mdms/downloads/MDMS-overview.pdf

13. Electromechanical meter and solid-state meter. http://en.wikipedia.org/wiki/Electric_energy_meter

14. M. Conner, *Sensors Empower the Internet of Things*, 2010, pp. 32–38.

15. MuNet Meters, 2009. http://www.munet.com/.

16. Aclara Software, *Meter Data Management: The Key to Unlocking the Benefit of Advanced Metering*, Aclara Software White Paper. Hazelwood, MO: Aclara, March 2008.

17. American National Standards Institute, *ANSI C12.18-2006, American National Standard Protocol Specification for ANSI Type 2 Optical Port.* New York: American National Standards Institute, 2006.

18. American National Standards Institute, *ANSI C12.21-2006, American National Standard Protocol Specification for Telephone Mode.* New York: American National Standards Institute, 2006.

19. International Organization for Standardization/International Electrotechnical Commission, *ISO/IEC Standard 7498-1:1994.*

20. International Electrotechnical Commission, IEC 62056 workshop in New Delhi, February 2009. http://www.dlms.com/news/0000009c300e1ae01.html

21. International Electrotechnical Commission, *IEC 62056-62 the Interface Class for Electricity Metering Data Exchange for Meter Reading, Tariff and Load Control*, 2nd ed. Geneva, Switzerland: IEC, 2006.

22. American National Standards Institute, *ANSI C12.19-2008, American National Standard—Utility Industry End Device Data Tables*, approved February 24, 2009. New York: American National Standards Institute.

23. Itron. *The AMI/AMR Solution from Itron Inc.* http://www.itron.com/pages/products_category.asp?id=itr_000238.xml

24. Elster Electricity, *EnergyAxis from Elster Electricity LLC.* http://www.elsterelectricity.com/internet_Content_1.nsf/SResults/D72B4A78CC3B0A1B85256DFF006EF2C3

25. Trilliant, *Trilliant—A Trusted Solution Partner*, Solution Brief, Trilliant Incorporated, 2009. http://www.trilliantinc.com/4_Rsrcs/_PDFs/TSB_TrustedPartner.pdf

26. Y. T. Lee, *Information Modeling from Design to Implementation.* New York, National Institute of Standards and Technology, 1999.

27. International Electrotechnical Commission, *IEC 61968-9 Ed. 1 Part 9: Interface for Meter Reading and Control.* New York: IEC/TC 57, August 14, 2009.

28. G. A. McNaughton and B. Saint, Integration using the MultiSpeak® Specification, Paper presentation at *Utility Automation*, December 2008.

29. Z. Li, Z. Wang, et al., A unified solution for advanced metering infrastructure integration with a distribution management system, *2010 First IEEE Smart Grid Communication*, NIST, Washington, DC, October 2010, pp. 566–571.

30. Z. Wang and Z. Li, *Meter Data Integration for Distribution Management System*, Tech Report No. CRID80345&80596. Raleigh, NC: ABB U.S. Corporate Research Center, 2009.

31. Christian Nagel Enterprise Services with the .NET Framework, Microsoft.net Development Series, January 13, 2005.

32. Java Message Service. http://en.wikipedia.org/wiki/Java_Message_Service

33. IBM, *IBM WebSphere MQ*. http://www-01.ibm.com/software/integration/wmq/.

5

COGNITIVE RADIO NETWORK FOR THE SMART GRID

RAGHURAM RANGANATHAN, ROBERT QIU, ZHEN HU, SHUJIE HOU, ZHE CHEN, MARBIN PAZOS-REVILLA, AND NAN GUO

Contents

Recently, cognitive radios and the smart grid are two areas that have received considerable research impetus. Cognitive radios are fully programmable wireless devices that can sense their environment and dynamically adapt their transmission waveform, channel access method, spectrum use, and networking protocols. It is widely anticipated that cognitive radio technology will be used for a general-purpose programmable radio that will serve as a universal platform for wireless system development, much like microprocessors have served a similar role for computation. The salient features of the cognitive radio (i.e., frequency agility, transmission speed, and range) are ideal for application to the smart grid. In this regard, a cognitive radio network can serve as a robust and efficient communications infrastructure that can address both the current and future energy management needs of the smart grid. The cognitive radio network can be deployed as a large-scale wireless regional-area network (WRAN) in a smart grid to utilize the unused TV bands recently approved for use by the Federal Communications Commission (FCC). In addition, a cognitive radio network test bed for the smart grid would serve as an ideal platform not only to address various issues related to the smart grid (e.g., security, information flow and power flow management, etc.) but also to reveal more practical problems for further research. In this chapter, the novel concept of incorporating a cognitive radio network as the communications backbone for the smart grid is outlined. A brief overview of the cognitive radio is provided, including the recently proposed Institute of Electrical and Electronics Engineers (IEEE) 802.22 standard. In particular, an overview of the cognitive radio network test bed, existing and new hardware platforms for cognitive radio networks, and functional architectures is given. Cognitive machine learning approaches such as principal component analysis (PCA) and kernel PCA for dimensionality reduction of high-dimensional smart grid data are presented. In addition, a novel approach of combining the recently developed robust PCA algorithm with a statistical signal-processing method called independent component analysis (ICA) is described for recovery of smart meter wireless transmissions in the presence of strong wideband interference. Security for the smart grid is still in the incipient stages and is the topic of significant research focus. This chapter addresses the impending problem of securing the smart grid, in addition to the possibility of applying fuzzy logic intrusion detection based on field-programmable gate array (FPGA) for the smart grid.

5.1 Introduction

5.1.1 Cognitive Radio

Cognitive radio is an intelligent software-defined radio (SDR) technology that facilitates efficient, reliable, and dynamic use of the underused radio spectrum by reconfiguring its operating parameters and functionalities in real time depending on the radio environment. Cognitive radio networks promise to resolve the bandwidth scarcity problem by allowing unlicensed devices to transmit in unused "spectrum holes" in licensed bands without causing harmful interference to authorized users.[1-4] In concept, the cognitive technology configures the radio for different combinations of protocol, operating frequency, and waveform. Current research on cognitive radio covers a wide range of areas, including spectrum sensing, channel estimation, spectrum sharing, and medium access control (MAC).

Due to its versatility, cognitive radio networks are expected to be increasingly deployed in both the commercial and military sectors for dynamic spectrum management. To develop a standard for cognitive radios, the Institute of Electrical and Electronics Engineers (IEEE) 802.22 Working Group was formed in November 2004.[5] The corresponding IEEE 802.22 standard defines the physical (PHY) and MAC layers for a wireless regional-area network (WRAN) that uses white spaces within the television bands between 54 and 862 MHz, especially within rural areas where usage may be lower. Details of the IEEE 802.22 standard, including system topology, system capacity, and the projected coverage for the system are given in the next section.

5.1.2 The 802.22 System

IEEE 802.22 is the first standardized air interface for cognitive radio networks based on opportunistic utilization of the TV broadcast spectrum.[6,7] The main objective of the IEEE 802.22 standard is to provide broadband connectivity to remote areas with comparable performance to broadband technologies such as cable, DSL (digital subscriber loop), and so on in urban areas. In this regard, the FCC selected the predominantly unoccupied TV station channels operating in the VHF (very-high-frequency) and UHF (ultra-high-frequency) region of the radio spectrum.

5.1.2.1 System Topology The 802.22 system is a point-to-multipoint wireless air interface consisting of a base station (BS) that manages a cell comprised of a number of users or customer premises equipment (CPEs).[8] The BS controls the medium access and "cognitive functions" in its cell, transmits data to the CPEs in the downlink, while receiving data in the uplink direction from the CPEs. The various CPEs perform distributed sensing of the signal power in the assorted channels of the TV band. In this manner, the BS collects the different measurements from the CPEs and exploits the spatial diversity of the CPEs to make a decision if any portion of the spectrum is available.

5.1.2.2 Service Coverage Compared to other IEEE 802 standards, such as 802.11, the 802.22 BS coverage range can reach up to 100 km if not limited by power constraints. The coverage of different wireless

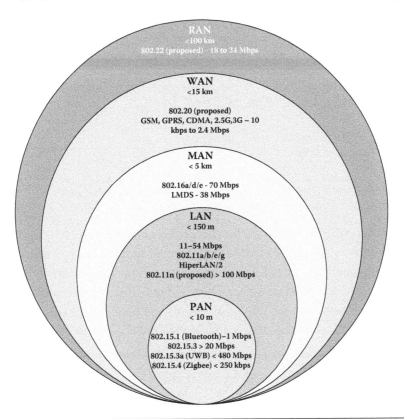

Figure 5.1 Comparison of 802.22 with other wireless standards. CDMA = code division multiple access; GPRS = general packet radio services; GSM = Global System for Mobile Communications; HiperLAN = High Performance Radio LAN; LAN = local-area network; LMDS = local multipoint distribution service; MAN = metropolitan-area network; PAN = personal area network; RAN = regional-area network; UWB = ultra-wideband; WAN = wide-area network.

standards is shown in Figure 5.1. The WRAN has the highest coverage due to higher transmit power and long-range propagation characteristics of TV bands.

5.1.2.3 System Capacity The WRAN systems can achieve comparable performance to that of DSL, with downlink speeds of 1.5 Mbps and uplink speed of 384 kbps. The system would thus be able to support 12 simultaneous CPEs, resulting in an overall system download capacity of 18 Mbps.

The specification parameters of the IEEE 802.22 standard are summarized in Table 5.1.

In Section 5.2, the concept of developing a cognitive radio network for the smart grid is presented, in addition to an overview of various

Table 5.1   IEEE 802.22 Characteristics

PARAMETER	SPECIFICATION
Typical cell radius (km)	30–100 km
Methodology	Spectrum sensing to identify free channels
Channel bandwidth (MHz)	6, 7, or 8
Modulation	OFDMA
Channel capacity	18 Mbps
User capacity	Downlink: 1.5 Mbps
	Uplink: 384 kbps

Source: From IEEE, with permission. OFDMA = Orthogonal Frequency-Division Multiple Access.

existing hardware platforms for cognitive radio networks. Section 5.3 outlines new approaches for the development of hardware test beds for smart grid cognitive radio networks. In Section 5.4, cognitive algorithms for preprocessing and recovery of high-dimensional smart grid data are illustrated. Section 5.5 addresses the critical issue of security in smart grid communications, followed by conclusions in Section 5.6.

5.2 Cognitive Radio Network for Smart Grid

The smart grid explores and exploits two-way communication technology, advanced sensing, metering and measurement technology, modern control theory, network grid technology, and machine learning in the power system to make the power network stable, secure, efficient, flexible, economical, and environmentally friendly. To support the smart grid, a dedicated two-way communications infrastructure should be set up for the power system. In this way, secure, reliable, and efficient communication and information exchange can be guaranteed. In addition, the various devices, equipment, and power generation facilities of the current power system should be updated and renovated. Novel technologies for power electronics should be used to build advanced power devices (e.g., transformer, relay, switch, storage, and so on).

In the area of wireless communications, cognitive radio is an emerging technique. The essence of cognitive radio is the ability of communicating over the unused frequency spectrum adaptively and intelligently. The idea of using cognitive radio in the smart grid appears to be proposed in the literature, for the first time, in Qiu[9–11]

Table 5.2 Advantages of Applying the Cognitive Radio (CR) to the Smart Grid

SALIENT FEATURES	DESCRIPTION
Frequency diversity	CR can operate over unused frequency bands
Transmission speed	Data rates of up to tens of megabits per second can be achieved
Range	CR can transmit over long distances in a WRAN scenario
Adaptability	CR has inherent intelligence to adapt to changes in the environment
Programmability	Built on an SDR platform, the CR can be selectively programmed

and Qiu et al.[12] The capability of cognitive radio enables the smart grid, in many aspects, including security. With minimal modifications to software, a cognitive radio network can be used for efficient control of the smart grid.

The benefits of applying cognitive radio to the smart grid are summarized in Table 5.2. First, cognitive radio can operate over a wide range of frequency bands. It has frequency agility. This feature is especially useful for the smart grid because the frequency spectrum today is so crowded, and cognitive radio provides the capability of reusing unused frequency bands for the smart grid. Second, cognitive radio enables high-speed data transmission for the smart grid. This is due to the wideband nature of cognitive radio. The data rate can be as high as tens of megabits per second, in contrast to the ZigBee, which can only provide a data rate of tens to hundreds of kilobits per second. Third, cognitive radio has the potential to transmit data over a long distance. Recently, the Federal Communications Commission (FCC) has decided to allow use of unused TV bands for wireless communications. The TV bands are ideal for long-distance mass data transmission. Cognitive radio in a WRAN scenario is designed to utilize the unused TV bands. Employing cognitive radio, the smart grid can communicate over a long distance over the air. Fourth, cognitive radio boasts of cognitive learning and adaptation capability. It has the ability to learn the environment, reason from it, and adapt accordingly. Cognitive radio makes the smart grid "smarter" and more robust. Fifth, cognitive radio is based on the SDR platform, which is a programmable radio. Hence, cognitive radio is capable of performing different applications and tasks. In addition, security, robustness, reliability, scalability, and sustainability of the smart grid can be effectively supported by cognitive radio due to its flexibility and reprogrammability.

5.2.1 Cognitive Radio Network Test Bed: Hardware Platforms for Cognitive Radio Networks

There have been some wireless network test beds, such as the Open Access Research Testbed for Next-Generation Wireless Networks (ORBIT)[13] and the wireless test bed developed by University of California, Riverside.[14] Some common features of those wireless network test beds are summarized as follows: First, the nodes in the networks are developed based on computer central processing units (CPUs). Second, the nodes use 802.11 Wi-Fi network interface cards for wireless communications. These network test beds may work well for evaluating algorithms, protocols, and network performances for Wi-Fi networks, but they are not suitable for cognitive radio networks due to their inherent lack of wideband frequency agility.

Recently, Virginia Tech developed a test bed for cognitive radio networks with 48 nodes,[15] which is a significant achievement in this area. Each node consists of three parts: an Intel Xeon processor-based high-performance server, a Universal Software Radio Peripheral 2 (USRP2), and a custom-developed radio-frequency (RF) daughterboard that covers a continuous frequency range from 100 MHz to 4 GHz with variable instantaneous bandwidths from 10 kHz to 20 MHz. The node is easily capable of frequency agility. However, as the authors mentioned, the drawbacks of the node are twofold. First, it is not a low-power processing platform. Second, it is not capable of mobility.

Regardless of the kind of cognitive radio network test bed, it is composed of multiple nodes. There exist some commercial off-the-shelf hardware platforms designed for SDR that may be used for building the nodes for cognitive radio networks.

5.2.1.1 Universal Software Radio Peripheral 2 USRP and USRP2, provided by Ettus Research, are widely used hardware platforms in the area of SDR and cognitive radio. USRP2 is the second generation of USRP, and it became available in 2009.[16] USRP2 consists of a motherboard and one or more selectable RF daughterboards, as shown in Figure 5.2.

The major computation power on the motherboard comes from a Xilinx Spartan-3 XC3S2000 field-programmable gate array (FPGA). The motherboard is also equipped with a 100-mega-samples per second (MSPS), 14-bit, dual-channel analog-to-digital converter (ADC); a

Figure 5.2   USRP2 with WBX RF daughterboard.

400-MSPS, 16-bit, dual-channel digital-to-analog converter (DAC); and a Gigabit Ethernet port that can be connected to a host computer. There are some RF daughterboards available for USRP2. Among them, a newly developed RF daughterboard called wide bandwidth transceiver (WBX) covers a wide frequency band of 50 MHz to 2.2 GHz, with a nominal noise figure of 5–7 dB.

Signals are received and downconverted by USRP2 and its RF daughterboard. Subsequently, they are sent to a host computer for further processing through the Gigabit Ethernet. Most of the processing work is done by the host computer. Data to be transmitted are sent from the host computer to USRP2 through the same Gigabit Ethernet before they are upconverted and transmitted by USRP2 and its RF daughterboard.

A major advantage of USRP2 is that it works with GNU Radio,[17] an open source software with plenty of resources for SDR and many users, which simplifies and eases the use of USRP2. On the other hand, USRP2 is not perfect. First, the Gigabit Ethernet connecting USRP2 and its host computer introduces random time delays. The operating system on the host computer may also introduce random time delays. According to our measurement, the response delay of USRP2 is in the range of several milliseconds to tens of milliseconds.[18] Such random response delay may be acceptable for half-duplex communications. However, in cognitive radio networks, full-duplex communications are desired, and random response delays may deteriorate the performance of cognitive radio networks. Second,

USRP2 is usually used together with GNU Radio that runs on a host computer. When the instantaneous bandwidth of USRP2 increases, the CPU on the host computer becomes much busier. Therefore, a multicore CPU is desired, similar to what Virginia Tech has done to its network test bed. When the instantaneous bandwidth of USPR2 becomes wider and the processing tasks on GNU Radio become much more complex, a common CPU may not be competent enough for real-time processing.

5.2.1.2 Small Form Factor Software-Defined Radio Development Platform
The small form factor (SFF) SDR development platform (DP) provided by Lyrtech in collaboration with Texas Instruments (TI) and Xilinx is a self-contained platform consisting of three separate boards: digital processing module, data conversion module, and RF module, as shown in Figure 5.3.[19–21]

The digital processing module is designed based on TMS320DM6446 System-on-Chip (SoC) from TI and Virtex-4 SX35 FPGA from Xilinx. The TMS320DM6446 SoC has a C64x+ digital signal processor (DSP) core running at 594 MHz together with an advanced reduced instruction set computing (RISC) machine (ARM9) core running at 297 MHz. The digital processing module also comes with a 10/100-Mbps Ethernet port. The data conversion module is equipped with a 125-MSPS, 14-bit, dual-channel ADC and a 500-MSPS, 16-bit, dual-channel DAC. It also has a Xilinx Virtex-4

Figure 5.3 SFF SDR DP with low-band tunable RF module.

LX25 FPGA. The low-band tunable RF module can be configured to have either 5- or 20-MHz bandwidth with working frequencies of 200–1,050 MHz for the transmitter and 200–1,000 MHz for the receiver. The nominal noise figure of this RF module is 5 dB. Other frequency bands may be covered by several other RF modules.

There are two favorable features of the SFF SDR DP for cognitive radio networks. One is that a SFF SDR DP is in SFF and can be moved easily. The other is that it is capable of supporting full-duplex communications. However, there are also two technical drawbacks of using it to build nodes for cognitive radio networks. One drawback is that its computing capacity is fixed, and it is not easy to upgrade to meet the demands of cognitive radio networks. The other drawback is the response time delay. According to our measurement, the response delay of an SFF SDR DP is about tens of milliseconds, and the delay is constant.[18] Such a nontrivial delay is undesirable for cognitive radio networks since it may deteriorate performance.

An SFF SDR DP can be viewed as an example of independent hardware platforms, whereas USRP2 is an example of computer-aided hardware platforms. A comparison between the two hardware platforms has been reported in Qiu et al.[12]

5.2.1.3 Wireless Open-Access Research Platform The Wireless Open-Access Research Platform (WARP) developed by Rice University consists of an FPGA board and one to four radio boards,[22] as shown in Figure 5.4. The second generation of the FPGA board has a Xilinx

Figure 5.4 WARP FPGA board with two radio boards.

Virtex-4 FX100 FPGA and a Gigabit Ethernet port.[23,24] The FPGA can be used to implement the physical layer of wireless communications. There are PowerPC processors embedded in the FX100 FPGA that can be used to implement MAC and network layers. The radio board incorporates a dual-channel, 65-MSPS, 14-bit ADC and a dual-channel, 125-MSPS, 16-bit DAC, covering two frequency ranges of 2,400–2,500 MHz and 4,900–5,875 MHz, with a bandwidth of up to 40 MHz.

The WARP platform is also an SFF independent hardware platform, which is attractive for building the nodes of cognitive radio networks. The second advantage of using WARP is that both the physical layer and MAC layer can be implemented on one FPGA, which may simplify the board design, compared to an "FPGA + DSP/ARM" architecture. Hence, time delays introduced by the interface between FPGA and DSP/ARM can be reduced. However, according to Mango Communications,[24] the Virtex-4 FPGA on WARP is not powerful enough to accommodate both transmitter and receiver functions at the same time. Thus, full-duplex communications desired by cognitive radio networks cannot be implemented using just one WARP platform.

5.2.1.4 Microsoft Research Software Radio Microsoft Research has developed a software radio (Sora) platform.[25] Sora is composed of a radio control board (RCB) and a selectable RF board, and it works with a multicore host computer. The RCB is shown in Figure 5.5.

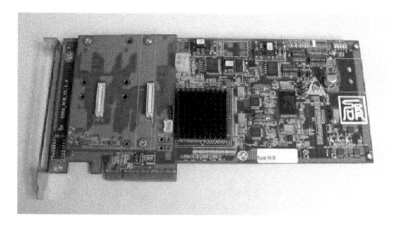

Figure 5.5 Sora radio control board.

The RCB contains a Xilinx Virtex-5 FPGA, and it interfaces with a host computer through a Peripheral Component Interconnect Express (PCIe) interface at a rate of up to 16.7 Gbps. Actually, the RCB is an interface board for transferring digital signals between the RF board and computer memory. The RF board can be a WARP radio board. Processing work, including physical layer and MAC layer, is done on the host computer.

Sora is a computer-aided platform. The main advantage of using Sora is that it provides a high-throughput interface between RF boards and a host computer. However, since processing work burdens the host computer, the host computer has to be very powerful to support all the functions running in real time. On the other hand, multicore programming and debugging with speed-up tricks is not easy. Moreover, implementing full-duplex communications on one host computer is challenging. Obviously, a host computer (or server) installed with Sora lacks mobility.

5.3 Innovative Test Bed for Cognitive Radio Networks and the Smart Grid

All of the four hardware platforms mentioned are designed for SDR. Two of them connect to a host computer where major processing work is done. The other two are stand-alone hardware platforms. From the aspect of mobility, stand-alone platforms are preferable for building the nodes of cognitive radio networks, whereas from the aspect of software development, computer-aided hardware platforms are more practical since software development and debugging on a host computer are generally easier. In Chowdhury and Melodia,[26] a compromise between the two kinds of hardware platforms is suggested. The authors recommended performing time-critical tasks in the FPGA and a split MAC design with host and FPGA implementations.

However, compared to the hardware platforms for SDR, the major concerns about hardware platforms for cognitive radio networks are computing power and response time delay. Cognitive radio introduces "intelligence" beyond SDR, like detection and learning algorithms, which means cognitive radio requires much more computing power than SDR. A hardware platform with ample and upgradable computing power is desired for building cognitive radio test beds. On the other hand, the desired hardware platform should have minimum response time delay. If the response time delay is large, the throughput of cognitive radio networks will seriously degrade. Moreover, full-duplex communications for the desired hardware platforms are preferable.

Unfortunately, none of the existing off-the-shelf hardware platforms can meet these requirements at the same time. They are originally designed for SDR instead of cognitive radio networks. It is imperative to design a new hardware platform for building the nodes of cognitive radio networks.

An innovative cognitive radio network test bed is being built at Tennessee Technological University.[12,27] The idea of applying a cognitive radio network test bed to the smart grid was developed there in the middle of 2009 in a funded research proposal.[28] Subsequently, this idea has been strengthened.[10,12,29–31] The objective of this test bed is to achieve the convergence of cognitive radio and the smart grid.[32]

The cognitive radio network test bed being built is unique and real-time oriented. It is designed to provide much more stand-alone computing power and reduce the response time delay. The cognitive radio network test bed is comprised of tens of nodes, with each node based on a self-designed motherboard, and commercial RF boards. On the self-designed motherboard, there are two advanced and powerful FPGAs that can be flexibly configured to implement any function. Therefore, this network test bed can be readily applied to the smart grid.

5.3.1 Motherboard for the New Hardware Platform

In this section, an architecture for the motherboard of the new hardware platform is given. Regarding the RF front end, existing RF boards from WARP or USRP2 can be reused to interface with this motherboard to constitute the new hardware platform.

Figure 5.6 shows the corresponding architecture of the first-generation new motherboard and its major components. Two powerful FPGAs (i.e., a Virtex-6 FPGA and a Virtex-5 FX FPGA) are employed as core components on the motherboard. All the functions for the physical and MAC layers are implemented on the two FPGAs, and no external host computer is required. This novel hardware platform stands alone; thus, it has good mobility. The Virtex-5

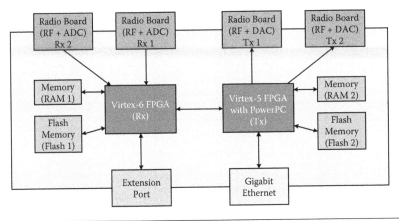

Figure 5.6 Architecture of the motherboard for the new hardware platform.

FX FPGA has PowerPC cores that are dedicated for implementing the MAC layer. Physical-layer functions, including spectrum sensing, are implemented on the two FPGAs. The Virtex-5 FPGA is used for the transmitting data path, and it is connected to one or two RF boards as well as a Gigabit Ethernet port. The Virtex-6 FPGA is dedicated for the receiving data path, with connections to one or two RF boards and an extension port. The extension port can be used to connect with external boards to gain access to additional computing resources. The two FPGAs are connected together by a high-throughput, low-latency onboard bus. Both of the FPGAs have access to their own external memories. The use of two FPGAs is a trade-off between performance and cost.

The new motherboard can provide enough and upgradable computing resources for cognitive radio networks. In addition, the time delays between the two FPGAs are trivial. Moreover, full-duplex communications are easily supported by this motherboard with two or more RF boards.

5.3.2 Functional Architecture for Building Nodes for Network Test Beds

Based on the new motherboard described in the previous section and off-the-shelf RF boards, nodes for network test beds can be implemented using the following functional architecture, as shown in Figure 5.7: The hardware abstraction layer (HAL) is a packaged interface for upper-level functions that screens hardware-specific details.

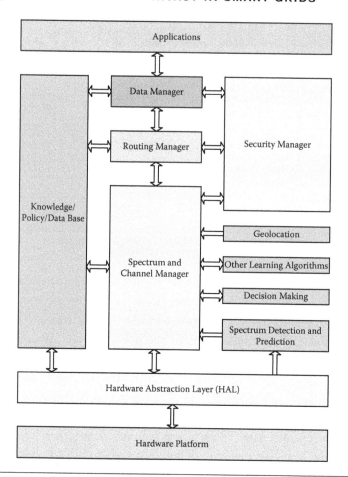

Figure 5.7   Functional architecture for the nodes.

It provides data interfaces to both receiving data and transmitting data paths, as well as an access interface to other hardware-specific resources on the hardware platform. The spectrum and channel manager manage all the spectrum- and channel-related resources, including links, frequencies, and modulation methods. There are several functional modules interfaced with the spectrum and channel manager. The spectrum detection and prediction module provides the information regarding the availability of some frequency bands. The decision-making module utilizes decision algorithms to make decisions such as which channel will be used and when it will be used. More learning algorithms can be implemented as an independent module to learn and reason from the inputs. The geolocation module outputs the latitude and longitude of the node. The spectrum and channel manager can use such geolocation information to load prior information about current location from the knowledge/policy/data database. The routing manager employs routing algorithms to select the best route for sending and relaying data packages. The data manager organizes all the data from upper-level applications and the data to be relayed. The security manager provides encryption and decryption to the data manager, routing manager, and spectrum and channel manager. The knowledge/policy/data database stores prior knowledge, policies, data, and experiences. After the nodes are built, a network test bed is ready to be established.

5.3.3 Innovative Network Test Bed

Multiple nodes constitute a network test bed. Figure 5.8 shows the innovative network test bed.

All the nodes are connected using Gigabit Ethernet to a console computer through an Ethernet switch. The console computer controls and coordinates all the nodes in the network test bed. This network test bed can be used not only for cognitive radio, but also for the smart grid. In smart grid applications, nodes of the network test bed implement microgrid central controllers, smart meters, or submeters. Adaptive wireless communications are incorporated into the nodes,

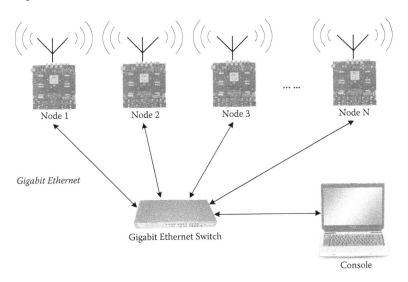

Figure 5.8 Innovative network test bed.

and information can be exchanged between microgrid central controllers, smart meters, and submeters.

5.4 Cognitive Algorithms for the Smart Grid

5.4.1 Dimensionality Reduction and High-Dimensional Data Processing in Cognitive Radio Networks

In cognitive radio networks, data exist in a significant amount. However, in practice, the data are highly correlated. This redundancy in the data increases the overhead of cognitive radio networks for data transmission and data processing. In addition, the number of degrees of freedom (DoF) in large-scale cognitive radio networks is limited. The DoF of a K user $M \times N$ multiple input multiple output (MIMO) interference channel has been discussed.[33] The total number of DoF is equal to $\min(M,N) * K$ if $K \le R$, and

$$\min(M,N) * \frac{R}{R+1} * K$$

if $K > R$, where

$$R = \frac{\max(M,N)}{\min(M,N)}.$$

This is achieved based on interference alignment.[34-36] Theoretical analysis about DoF in cognitive radio has been presented.[37,38] The DoF corresponds to the key variables or key features in the network. Processing the high-dimensional data instead of the key variables will not enhance the performance of the network. In some cases, this could even degrade the performance. Hence, compact representation of the data using dimensionality reduction is critical in cognitive radio networks.

5.4.1.1 Dimensionality Reduction Methods Dimensionality reduction[39-42] finds a low-dimensional embedding of high-dimensional data. Three dimensionality reduction methods can be employed— both linear methods such as principal component analysis (PCA)[43] and nonlinear methods such as kernel PCA (KPCA),[44] and landmark maximum variance unfolding (LMVU).[45,46] If we assume the original high-dimensional data as a set of M samples $\mathbf{x}_i \in \mathbf{R}^N$, $i = 1, 2,$ $\cdots$, M, then the reduced low-dimensional samples of $\mathbf{x}_i$ are $\mathbf{y}_i \in \mathbf{R}^K$, $i = 1, 2, \cdots , M$, where $K \ll N$. x_{ij} and y_{ij} are component-wise elements in $\mathbf{x}_i$ and $\mathbf{y}_i$, respectively.

PCA[43] is the best-known linear dimensionality reduction method; it performs linear mapping of the high-dimensional data to a low-dimensional space such that the variance of the low-dimensional data is maximized. In reality, the covariance matrix of the data is constructed, and the eigenvectors of this matrix are computed. The covariance matrix of $\mathbf{x}_i$ can be obtained as

$$\mathbf{C} = \frac{1}{M} \sum_{i=1}^{M} (\mathbf{x}_i - \mathbf{u})(\mathbf{x}_i - \mathbf{u})^T \tag{5.1}$$

where

$$\mathbf{u} = \frac{1}{M} \sum_{i=1}^{M} \mathbf{x}_i$$

is the mean of the given samples, and T denotes the transpose operator.

The eigenvectors corresponding to the largest eigenvalues can be exploited to obtain a large portion of the variance of the original data. The original high-dimensional space can be reduced to a space spanned by a few dominant eigenvectors. PCA works well for the high-dimensional data with linear relationships but always fails in a nonlinear scenario. PCA can be applied in the nonlinear situation by a kernel,[47–50] called KPCA.[44] KPCA is therefore a kernel-based machine learning algorithm. It uses the kernel function, which is the same as the support vector machine (SVM), to implicitly map the original data to a feature space F where PCA can be applied.

Other nonlinear techniques for dimensionality reduction include manifold learning techniques. Within the framework of manifold learning, the current trend is to learn the kernel using semidefinite programming (SDP)[51–55] instead of defining a fixed kernel. The most prominent example of such a technique is MVU (maximum variance unfolding).[45] MVU can learn the inner product matrix of $\mathbf{y}_i$ automatically by maximizing their variance, subject to the constraints that $\mathbf{y}_i$ are centered and local distances of $\mathbf{y}_i$ are equal to the local distances of $\mathbf{x}_i$.

Here, the local distances represent the distances between y_i (x_i) and its k nearest neighbors, in which k is a parameter. The corresponding SDP can be cast into the following form:[45]

$$\text{maximize trace } (\mathbf{I})$$

$$\text{subject to}$$

$$\mathbf{I} \succ 0$$

$$\sum_{ij} \mathbf{I}_{ij} = 0 \tag{5.2}$$

$$\mathbf{I}_{ii} - 2\mathbf{I}_{ij} + \mathbf{I}_{jj} = D_{ij}, \text{ when } \eta_{ij} = 1$$

where $\mathbf{I}$ is an inner product matrix of y_i, $D_{ij} = ||\mathbf{x}_i - \mathbf{x}_j||^2$, and $\mathbf{I} \succ 0$ implies that $\mathbf{I}$ is a positive semidefinite (PSD) matrix.

LMVU[46] is a modified version of MVU that aims to solve problems on a larger scale compared to MVU. It uses the inner product matrix $\mathbf{A}$ of randomly chosen landmarks from $\mathbf{x}_i$[46] to approximate the full matrix $\mathbf{I}$, in which the size of $\mathbf{A}$ is much smaller than $\mathbf{I}$. In this way, the speed of computing is increased.

5.4.1.2 Spectrum Monitoring Using Dimensionality Reduction and Support Vector Machine with Experimental Validation   Spectrum monitoring is one of the most challenging and critical tasks in cognitive radio networks. In this section, the feasibility of applying dimensionality reduction to the cognitive radio network is studied by presenting an experimental validation. The preliminary results[56] illustrate how to extract the intrinsic dimensionality of Wi-Fi signals by recent breakthroughs in dimensionality reduction techniques. This is a new trend in cognitive radio networks for spectrum monitoring, which differs from traditional spectrum-sensing techniques such as energy detection, matched filter detection, and cyclo-stationary feature detection.[57–59]

Wi-Fi time domain signals have been measured and recorded using an advanced digital phosphor oscilloscope (DPO), a Tektronix DPO72004.[60] The DPO supports a maximum bandwidth of 20 GHz and a maximum sampling rate of 50 GS/s. It is capable of recording up to 250 M samples per channel. In the measurements, a laptop accesses the Internet through a wireless Wi-Fi router, as shown in Figure 5.9. An antenna with a frequency range of 800 to 2,500 MHz is placed near the laptop and connected to the DPO. The sampling rate of the

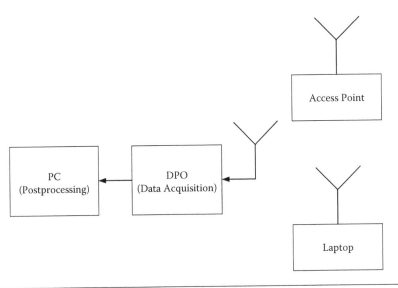

Figure 5.9 Setup for the measurement of Wi-Fi signals.

DPO is set to 6.25 GS/s. Recorded time domain Wi-Fi signals are shown in Figure 5.10. The duration of the recorded Wi-Fi signals is 40 ms.

The recorded 40-ms Wi-Fi signals are divided into 8,000 slots, with each slot lasting 5 μs. These slots can be viewed as spectrum-sensing slots. The time domain Wi-Fi signals within the first 1 μs of every slot are then transformed into the frequency domain using the fast Fourier transform (FFT), which is equivalent to FFT-based spectrum sensing. The frequency band of 2.411–2.433 GHz is considered. The resolution in the frequency domain is 1 MHz. Therefore, for each slot, 23 points in the frequency domain can be obtained, of which 13 points will be selected in the following experiment.

SVM is exploited to classify the states (busy $l_i = 1$ or idle $l_i = 0$) of the measured Wi-Fi data with or without dimensionality reduction, given the true states. SVM will classify the states of the spectrum data at different time slots.

The DoF of the Wi-Fi frequency domain signals is extracted from the original 13 dimensions. The flowchart of the SVM processing combined with dimensionality reduction methods is shown in Figure 5.11. The false alarm rate obtained by combining SVM with dimensionality reduction and employing only SVM is shown in Figure 5.12.

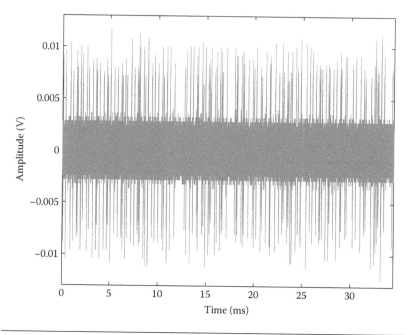

Figure 5.10   Recorded Wi-Fi signals in time domain.

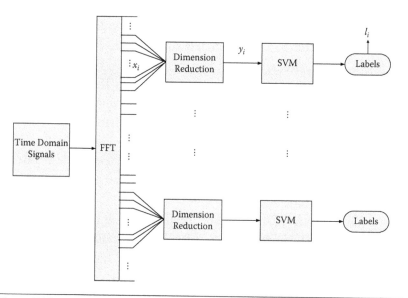

Figure 5.11   The flowchart of SVM combined with dimensionality reduction.

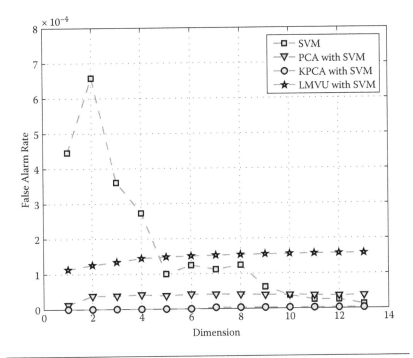

Figure 5.12 False alarm rate.

The original dimension of the frequency domain data varies from 1 to 13 for the SVM method. In addition, the SVM method is applied to the data with the extracted dimensions from 1 to 13, obtained by dimensionality reduction.

Experimental results showed that with dimensionality reduction, the performance was much better than that without dimensionality reduction.

5.4.2 Robust Principal Component Analysis

In many practical problems, the collected data can be organized in matrix form. Usually, the size of the matrix is huge. However, the DoF of the matrix are finite, which means the matrix is low rank.

A well-known low-rank matrix approximation algorithm is PCA.[61] If the observation matrix is **R**, PCA finds a low-rank approximation of the original matrix **R** by solving the optimization model

$$\min_{\mathbf{L}} \|\mathbf{R} - \mathbf{L}\|, \text{ subject to } \text{rank}(\mathbf{L}) \leq r \qquad (5.3)$$

in which $\|\cdot\|$ is the spectral norm of a matrix (the largest singular value of the matrix). PCA finds the optimal low-rank approximation in the least-square sense. This problem can be simply solved by singular value decomposition (SVD). However, an intrinsic drawback of PCA is that it can work efficiently only when the low-rank matrix is corrupted with independent and identically distributed (i.i.d.) Gaussian noise. That is, PCA is suitable for the model of

$$\mathbf{R} = \mathbf{L} + \mathbf{N} \tag{5.4}$$

in which $\mathbf{L}$ is the low-rank matrix, and $\mathbf{N}$ is the i.i.d. Gaussian noise matrix. However, it will fail when some of the entries in $\mathbf{L}$ are grossly corrupted,

$$\mathbf{R} = \mathbf{L} + \mathbf{S} \tag{5.5}$$

in which $\mathbf{L}$ is still the low-rank matrix, but the matrix $\mathbf{S}$ is a sparse matrix with arbitrarily large magnitude, and the number of nonzero entries is m.

The problem of recovering the low-rank matrix from a grossly corrupted observation matrix has been solved efficiently by the relaxed convex optimization model (principal component pursuit):[62]

$$\min_{\mathbf{L},\mathbf{S}} \|\mathbf{L}\|_* + \lambda \|\mathbf{S}\|_1, \text{subject to } \mathbf{R} = \mathbf{L} + \mathbf{S}, \tag{5.6}$$

in which $\|\cdot\|_*$ represents the nuclear norm of a matrix (sum of the singular values), $\|\cdot\|_1$ denotes the sum of the absolute values of matrix entries, and λ is a trade-off parameter. It has been thoroughly investigated[62,63] that as long as $\mathbf{S}$ is sparse enough, the formulated optimization problem (5.6) can exactly recover the low-rank matrix $\mathbf{L}$. This kind of problem has been traditionally called robust PCA,[62–64] which is closely related to, but harder than, the famous problem of matrix completion.[65–70]

One of the requirements for robust PCA is that the low-rank matrix cannot be sparse at the same time. An incoherence condition defined in Candès and Tao[65] and Candès and Recht[66] with parameter μ states that the singular vectors of $\mathbf{L}$ satisfy the following two assumptions:

$$\max_i \left\|\mathbf{U}^H \mathbf{e}_i\right\|^2 \leq \frac{\mu r}{M}, \quad \max_i \left\|\mathbf{V}^H \mathbf{e}_i\right\|^2 \leq \frac{\mu r}{L} \tag{5.7}$$

and

$$\left\| \mathbf{UV}^H \right\|_{\infty} \le \sqrt{\frac{ur}{ML}} \tag{5.8}$$

where $\|\cdot\|_{\infty}$ is the maximum absolute value of all the entries in the matrix, H denotes conjugate transpose, and $\mathbf{e}_i$ is the canonical basis vector in Euclidean space. The matrices are $\mathbf{U} = [\mathbf{u}_1, \mathbf{u}_2, \cdots, \mathbf{u}_r]$ and $\mathbf{V} = [\mathbf{v}_1, \mathbf{v}_2, \cdots, \mathbf{v}_r]$. $\mathbf{u}_i$, $i = 1, 2, \cdots, r$ and $\mathbf{v}_i$, $i = 1, 2, \cdots, r$ are the left and right singular vectors obtained by performing SVD on $\mathbf{L}$:

$$\mathbf{L} = \sum_{i=1}^{r} \sigma_i \mathbf{u}_i \mathbf{v}_i^H, \tag{5.9}$$

where σ_i, $i = 1, 2, \cdots, r$ are positive singular values, and $\mathbf{L}$ is a rank r matrix with size $M \times L$. The incoherence condition implies that the entries in the singular vectors $\mathbf{u}_i$, $i = 1, 2, \cdots, r$ and $\mathbf{v}_i$, $i = 1, 2, \cdots, r$ are spread out.

A theorem based on the two assumptions in (5.7) and (5.8) has been proposed and proved[62] and is stated as follows:

Theorem 1.[62] Suppose $\mathbf{L}$ is a rectangular matrix of size $M \times L$; there is a numerical constant c such that principal component pursuit with $\lambda = 1/\sqrt{M_{(1)}}$ succeeds with probability at least $1 - cM_{(1)}^{-10}$, provided that

$$\mathrm{rank}(\mathbf{L}) \le \rho_r M_{(2)} \mu^{-1} (\log M_{(1)})^{-2} \tag{5.10}$$

$$m \le \rho_s ML, \tag{5.11}$$

The matrix $\mathbf{L}$ obeys (5.7) and (5.8), and the support set of $\mathbf{S}$ is uniformly distributed among all sets of cardinality m, in which $M_{(1)} = \max(M, L)$, $M_{(2)} = \min(M, L)$; ρ_r and ρ_s are positive numerical constants.

The theorem states that the low-rank matrix $\mathbf{L}$ and sparse matrix $\mathbf{S}$ (with arbitrarily large magnitude) can be exactly recovered from the observation matrix $\mathbf{R} = \mathbf{L} + \mathbf{S}$ with very large probability once the assumptions of the theorem are satisfied, that is, $\hat{\mathbf{L}} = \mathbf{L}$ and $\hat{\mathbf{S}} = \mathbf{S}$ are exact. The original low-rank and sparse matrices are expressed by $\mathbf{L}$ and $\mathbf{S}$, respectively. The recovered (extracted) low-rank and sparse matrices are expressed by $\hat{\mathbf{L}}$ and $\hat{\mathbf{S}}$, respectively.

In the presented simulations, the inexact augmented Lagrange multiplier (IALM)[71] method is employed to recover the sparse component $\hat{\mathbf{S}}$ and the low-rank component $\hat{\mathbf{L}}$ from the observation matrix $\mathbf{R}$. The parameters for the IALM algorithm are set identical to the default values of the code, which can be downloaded from the Web site.[72] The errors between the recovered and the original matrices are computed by

$$\frac{\left\|\hat{\mathbf{L}}-\mathbf{L}\right\|_F}{\left\|\mathbf{L}\right\|_F}, \quad \frac{\left\|\hat{\mathbf{S}}-\mathbf{S}\right\|_F}{\left\|\mathbf{S}\right\|_F}. \tag{5.12}$$

The simulation results are based on the theoretical covariance matrix of a random process

$$y(n) = x(n) + w(n), \tag{5.13}$$

in which

$$x(n) = \sum_{l=1}^{L} A_l \sin(2\pi f_l nT + \theta_l), \tag{5.14}$$

$x(n)$ and $w(n)$ are assumed to be independent, and $w(n)$ is added zero-mean white noise.

The Mth order covariance matrix of this process is

$$\mathbf{R}_{yy} = \mathbf{R}_{xx} + \sigma^2 \mathbf{I}, \tag{5.15}$$

where $\sigma^2 \mathbf{I}$ denotes the covariance matrix of noise with power spectral density σ^2 and $\mathbf{R}_{xx}$ denotes the covariance matrix of random signal. $\mathbf{I}$ represents the Mth order identity matrix.

The Mth order covariance matrix for $x(n)$ can be written as[73]

$$\mathbf{R}_{xx} = \sum_{l=1}^{L} \frac{A_l^2}{4} \left[\mathbf{e}_M(f_l)\mathbf{e}_M^H(f_l) + \mathbf{e}_M^*(f_l)\mathbf{e}_M^T(f_l) \right] \tag{5.16}$$

where H denotes complex conjugate transposition, $*$ denotes complex conjugation, and

$$\mathbf{e}_M(f_l) = \begin{pmatrix} 1 \\ \exp(j2\pi f_1 T) \\ \vdots \\ \exp(j2\pi f_1 MT) \end{pmatrix}. \tag{5.17}$$

The rank of matrix (5.16) is $2L$.

From (5.15), the theoretical covariance matrix $\mathbf{R}_{yy}$, which is the observation matrix $\mathbf{R}$ here, is comprised of the sparse component $\sigma^2 \mathbf{I}$ expressed by $\mathbf{S}$ and low-rank component $\mathbf{R}_{xx}$ expressed by $\mathbf{L}$ with rank $2L$. Robust PCA can be explored to separate the low-rank and sparse components from observation matrix $\mathbf{R}$.

First, considering the case of $L = 1$, $A_l = 1$, $f_l = 0.02l$, $T = 1$ of (5.14), and the order of covariance matrix $M = 128$, the results obtained by applying the IALM algorithm to the matrix $\mathbf{R}_{yy}$ are shown in Figure 5.13.

Corresponding results achieved by applying the IALM algorithm to the matrix $\mathbf{R}_{yy}$ of $L = 3$, $A_l = 1$, $f_l = 0.02l$, $T = 1$ of (5.14) and the order of covariance matrix $M = 128$ are shown in Figure 5.14.

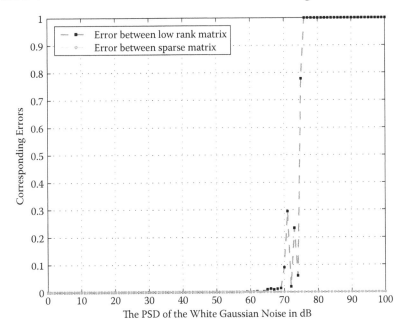

Figure 5.13 Errors between extracted and original matrices of one real sinusoidal function.

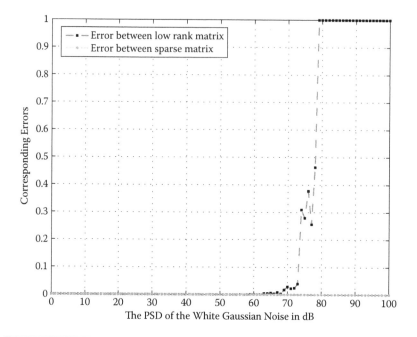

Figure 5.14 Errors between extracted and original matrices of three real sinusoidal functions.

Based on Figures 5.13 and 5.14, it can be seen that even if the power spectral density of white noise increases to 70 dB (approximated value), the IALM algorithm can still separate the low-rank and sparse components from the observation matrix $\mathbf{R}$ successfully via theoretical analysis.

In the next section, the robust PCA algorithm is employed as a preprocessing technique to mitigate strong wideband interference before applying the ICA approach for recovering the wireless smart meter transmissions.

5.4.3 Independent Component Analysis with Robust PCA Preprocessing for Recovery of Smart Meter Wireless Transmissions in the Presence of Strong Wideband Interference

Smart meters form an integral part of the smart grid. A smart meter is an electrical meter that records power consumption at regular intervals and communicates, either through power line communications or wireless transmissions, that information to the utility company for monitoring and billing purposes. Since the vision of a wireless cognitive radio network for the smart grid is presented in this chapter, smart meters equipped with wireless transmitters are considered. In this regard, the concept of ICA in combination with the robust PCA technique is presented as a possible approach to recover the simultaneous smart meter wireless transmissions in the presence of strong wideband interference.

5.4.3.1 *Independent Component Analysis Signal Model and Receiver Block Diagram* Independent component analysis is a statistical signal-processing method for extracting underlying independent components from multidimensional data,[74–77] In Liao and Niebur,[78] ICA was also applied to load profile estimation in electric transmission networks. ICA is very closely related to the method called blind source separation (BSS) or blind signal separation.[79–81] The term *blind* refers to the fact that we have little or no knowledge about the system that induces mixing of the source signals.

In a smart meter network, it is critical to accurately recover the smart meter wireless transmissions at the central node or access point (AP). In achieving this objective, one of the foremost challenges is the robustness of the data recovery in the presence of strong wideband interference due to easy access of the wireless data to unauthorized personnel and inadequacy of existing physical-layer security measures. In this section, a blind estimation approach to smart meter data recovery is presented by applying a complex ICA technique[82] in combination with the recently developed robust PCA algorithm[62] for interference mitigation and security enhancement.

In a smart meter network, each smart meter measures the current load at regular intervals and conveys that information to the control center at the power utility station. In this section, a wireless smart meter network is assumed in which each smart meter is equipped with a wireless transmitter, and the AP at the power utility control center collects all the wireless transmissions for processing the information. Since an ICA-based algorithm is used for recovery of the wireless smart meter data, the smart meters can transmit their information simultaneously. In Husheng et al.,[83] the concept of compressed sensing[84,85] was exploited to recover the sparse smart meter data transmissions by applying the basis pursuit algorithm.[86] However, in Husheng et al., it was assumed that the AP has accurate knowledge of the channel flat fading parameters from the channel estimation period of the data frame. In this section, an ICA-based blind estimation approach is applied by exploiting the statistical properties of the source signals. As a result, channel estimation in each data frame can be avoided, thereby allowing more information to be sent in each frame. Furthermore, to enhance the security of transmitted data, recovery of the wireless smart meter transmissions in the presence of strong wideband interference is also considered. In this regard, the recently developed method of robust PCA can be used.[62,71] The robust PCA method exploits the low-rank and sparseness property of the autocorrelation matrices of the smart meter signal and wideband interferer, respectively, to effectively separate them prior to ICA processing.

The smart meter network is assumed to consist of N smart meters controlled by an AP, similar to the illustration given in Husheng et al.[83] The channel parameters are assumed to be static over the transmission period, with Rayleigh flat fading characteristics. The data transmission section in the frame is divided into several time slots during which the active smart meters can simultaneously transmit their readings. Mathematically, the signal matrix Z received by the AP can be expressed as the following linear ICA signal model:

$$Z = HPX + W \qquad (5.18)$$

H is the Rayleigh flat fading channel matrix between the meters and the AP, P is the pseudorandom spreading code matrix for the meters, X is the source signal matrix transmitted by the meters, and W is the additive white Gaussian noise (AWGN). The spreading code is known only to the AP and meters and is unique for each meter. Replacing HP by the matrix A, (5.18) becomes

$$Z = AX + W \qquad (5.19)$$

In the context of ICA, A is called the mixing matrix. The objective of ICA is to recover X by estimating a matrix $\tilde{A}$ that approximates the inverse of A. Subsequently, an estimate of the source signal matrix $\tilde{X}$ can be obtained, as given by the following equation:

$$\tilde{X} = \tilde{A}Z \qquad (5.20)$$

In contrast to the popular carrier sense multiple access (CSMA) protocol, which uses a random back-off to avoid collisions in transmissions, the significant advantage of employing an ICA-based approach is that it enables simultaneous transmission for the smart meters. This eliminates the problem of incurring significant delay in data recovery. Furthermore, since ICA is a "blind" estimator, it does not need any prior knowledge of the channel or the pseudorandom noise (PN) code matrix. As long as the smart meter transmissions are independent, which is always the case since the meters are spatially separated, ICA can exactly recover all the smart meter signals.

In this section, smart meter data recovery in the presence of strong wideband interference is also addressed. Hence, in the event of strong interference, (5.19) becomes

$$Z = AX + W + Y \tag{5.21}$$

Since Y is not part of the signal mixing model AX, ICA algorithms cannot recover the source signals X in the presence of the interferer. Hence, it is imperative to separate Y from the observation matrix Z before any ICA method can be applied. To accomplish this, the second-order statistics of the signal and interferer are exploited. In particular, the autocorrelation function of each row of Z is computed. Rewriting (5.21) in terms of the autocorrelation matrices, we obtain

$$R = L + S + E \tag{5.22}$$

In (5.22), L is the low-rank autocorrelation matrix of the signal mixture, S is the sparse autocorrelation matrix of the wideband interferer consisting of only diagonal entries, and E is the autocorrelation matrix of the AWGN component. Therefore, (5.22) can be written as

$$R = L + \sigma_{int}^2 I + E \tag{5.23}$$

where σ_{int} is the power of the interferer, and I is the identity matrix. In this manner, (5.22) exactly fits the robust PCA matrix model described in the previous section.[62] Therefore, the robust PCA technique can be readily applied to recover the low-rank signal autocorrelation matrix from the sparse interferer autocorrelation matrix. This procedure is repeated for all the rows of the observation matrix Z. Therein, once the interferer Y is separated from Z, the signal model becomes similar to (5.19), and ICA can be applied to recover the source signals or smart meter transmissions X.

The baseband block diagram of the ICA-based receiver (central node or AP) is shown in Figure 5.15. The various stages of a typical receiver, such as downconversion, analog-to-digital conversion, synchronization, and so on, are assumed to be completed prior to the data recovery stage in the illustrated receiver.

5.4.4 Simulation Results Using the Robust PCA-ICA Approach

Typically, in a smart meter network, only a few meters would be actively transmitting their data. As a result, the sparsity of the smart meter data transmission to the central processing node or AP was exploited[83] for applying the principle of compressed sensing. In this section, it is assumed that in a smart meter network, $N = 10$ meters are simultaneously transmitting in quadrature phase shift keying (QPSK) modulation format. As a result of the transmitted data being complex valued, a complex FastICA separation algorithm with a saddle point test called FicaCPLX[82] is used for the blind recovery of source signals. Since ICA is a block-based technique, the processing block length (number of columns of Z) is assumed to be 1,000 symbols. The performance of the robust PCA-ICA approach is studied for different values of σ_{int}^2 from 1 to 5. The signal-to-noise ratio (SNR) is set at 20 dB. The signal-to-interference ratio (SIR)[87] is used as the measure of performance and is given by the following equation:

$$
SIR = \frac{1}{2N} \sum_m \left(\sum_n \frac{|p_{mn}|^2}{\left(\max|P_m|^2\right)} - 1 \right) +
$$

$$
\frac{1}{2N} \sum_n \left(\sum_m \frac{|p_{mn}|^2}{\left(\max|P_n|^2\right)} - 1 \right)
$$

(5.24)

where $P = \tilde{A}A$ is the permutation matrix of order N, in our case, a 10×10 matrix. Here, $\max |P_m|$ and $\max |P_n|$ are the absolute maximum values of the mth row and nth columns of P, respectively. Ideally, P should be a permutation matrix consisting of only ones. However, due to the amplitude ambiguity introduced by the ICA technique, the recovered signals have to be scaled accordingly. This can be accomplished by including a small preamble at the beginning of

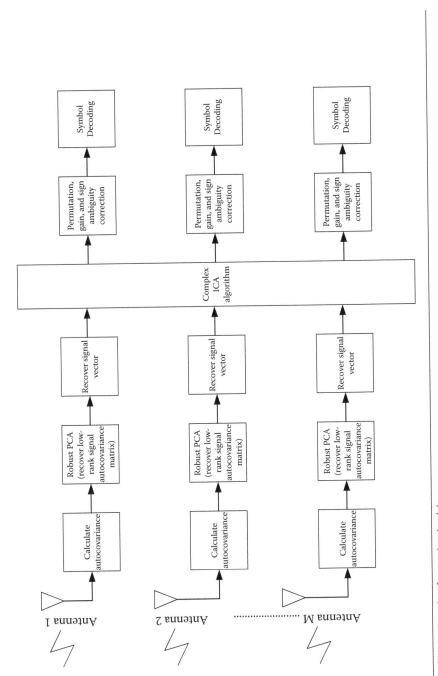

Figure 5.15 ICA-based receiver for smart meter data recovery.

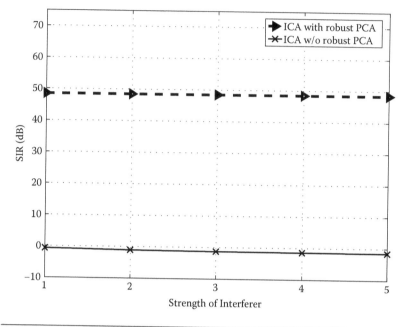

Figure 5.16   SIR(dB) versus σ^2_{int} for QPSK modulation.

each frame. The SIR (dB) achieved by the ICA algorithm FicaCPLX, with and without the robust PCA method for different σ^2_{int}, is shown in Figure 5.16. The constellation plots for the smart meter 1 QPSK signal before and after applying the FicaCPLX algorithm are shown in Figures 5.17 and 5.18, respectively.

5.5 Secure Communications in the Smart Grid

The smart grid is aimed at transforming the already-aging electric power grid in the United States into a digitally advanced and decentralized infrastructure with heavy reliance on control, energy distribution, communication, and security. Among the five identified key technology areas in the smart grid, the implementation of integrated communications is a foundational need.[88] The smart grid in the near future will be required to accommodate increased demands for improved quality and energy efficiency. Solar and wind farms are joining in for power generation in a distributed fashion. Appliances will become smart and talk to the control centers for optimum operations. Monitoring, managing, and controlling will be required at all levels. Prediction of electricity prices, weather, and social/human

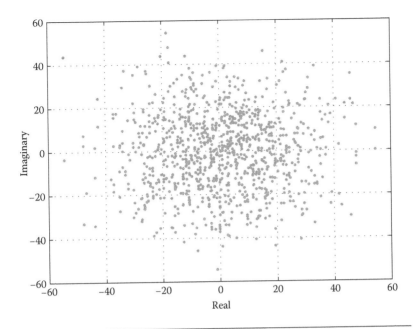

Figure 5.17 QPSK scatterplot before applying ICA.

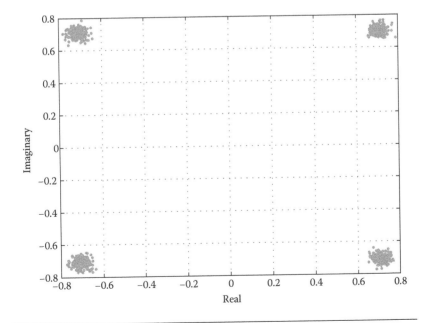

Figure 5.18 QPSK scatterplot after applying ICA.

activities will be taken into account for optimum control. The addition of these new elements will result in continuously increasing complexity. For different subnetworks or elements to be integrated into the smart grid seamlessly, a communication backbone has to be developed prior to adding various functions. Hence, the earlier the communication backbone is determined, the fewer the complications that will be faced later in building the grid.

5.5.1 Development of Communications Infrastructure

To develop this communications infrastructure, a high level of interconnectivity and reliability among its nodes is required. Sensors, advanced metering devices, electrical appliances, and monitoring devices, just to mention a few, will be highly interconnected, allowing for the seamless flow of data. Reliability and security in this flow of data between nodes, as shown in Figure 5.19, is crucial due to the low latency and cyberattack resilience requirements of the smart grid.

A distributed interconnection among these nodes will be ubiquitous, just as finding a similar level of connectivity among cellular phones or computing nodes in a large organization. The smart grid environment, however, poses a new set of communications and security paradigms. Due to their complexity and importance to the realization of the smart grid infrastructure, it is extremely important to

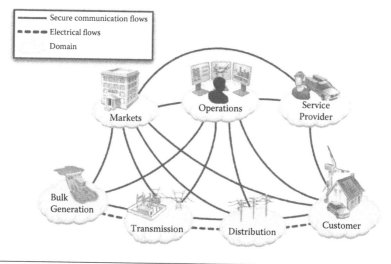

Figure 5.19 Interaction among actors in smart grid domains through secure communication flows and flows of electricity.

study the interactions among the nodes, more specifically, in terms of their communications and security.

Taking into account that reliability and security will impose constraints on the majority of the devices connected to the smart grid, if not all, it would be wise to consider communication standards, protocols, and devices that are designed from the ground up to be secured, logically and physically. Since a great portion of the traffic generated within the grid will be traveling on an unsecured medium such as the Internet, it is imperative to minimize the amount of potential security loopholes. In addition, the human variable should also be taken into account in the security model as part of the security infrastructure.

When it comes to security, communication is key, and information should be properly disseminated to all the parties involved, ensuring that everyone has a clear and common understanding of security needs facilitating their implementation and operation. Training and informing users about processes, study of human behavior, and the perception of events related to the processes are as important to the entire security equation as it is to engineer a secured infrastructure. As a matter of fact, the greatest security threat to any infrastructure is human error, as opposed to the technology securing it. Communications in the smart grid is a key component of the entire infrastructure, and logically we divide it into two sections: the backbone communications (interdomain), which will carry communications among domains such as those shown in Figure 5.19, and the communications at the LAN (intradomain) limited by perimeters such as a customer's house or a distribution facility.[89]

We can say that current and emerging technologies in telecommunications, most of which are expected to fall in the wireless realm (WiMAX, ZigBee, 802.11, etc.), can accommodate the communications needs of both inter- and intradomain environments, however, not without flaws. From a security standpoint, these technologies are not designed to be secure from the ground up. For example, ZigBee is a standard for short-range communications, and manufacturers of ZigBee-compliant chips produce them without necessarily considering the security issue. In addition, chip manufacturers print the chip model on top of the chip itself as a standard practice. The chip specifications can therefore be easily downloaded, and potential flaws of the chip can be easily exploited by attackers. Also, by default, many of these chips do not carry any internal security features and therefore rely on external chips or on higher-level software applications for this purpose. An easy access to the external chip by any malicious attacker could potentially disable any installed security features. This and other similar scenarios leads us to think that the smart grid should be driven by technologies and standards that consider security as their primary concern.

The smart grid has been conceived as being distributed in nature and heavily dependent on wireless communications. Today's SOHO (small office/home office) and enterprise-graded wireless devices include security features to mitigate attacks, with the vast majority still relying on conventional rule-based detection. It has been shown that conventional rule-based detection systems, although helpful, do not have the capability of detecting unknown attacks. Furthermore, as presented in Pazos-Revilla and Siraj,[90] these conventional intrusion detection systems (IDSs) would not be able to detect such an attack if it is carefully crafted since the majority of these rules are solely based on strict thresholds.

5.5.2 FPGA-Based Fuzzy Logic Intrusion Detection for the Smart Grid

Artificial intelligence techniques such as fuzzy logic, Bayesian inference, neural networks, and other methods can be employed to enhance the security gaps in conventional IDSs. As shown in Figure 5.20,

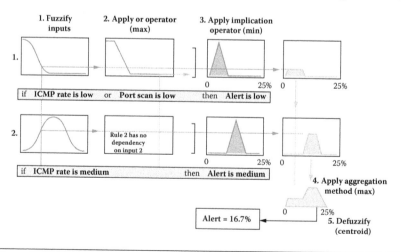

Figure 5.20   Fuzzy logic example applied to IDS. ICMP = Internet Message Control Protocol.

a fuzzy logic approach was used[91] in which different variables that influence the inference of an attack can be analyzed and later combined for the decision-making process of a security device. In addition, if each security device serving as an IDS is aware not only of itself but also of a limited number (depending on local resources and traffic) of surrounding trusted IDS devices, the alerts that these other devices generate can be used to adjust local variables or parameters to better cope with distributed attacks and more accurately detect their presence.

The research and development of robust and secure communication protocols, dynamic spectrum sensing, as well as distributed and collaborative security should be considered as an inherent part of smart grid architecture. An advanced decentralized and secure infrastructure needs to be developed with two-way capabilities for communicating information and controlling equipment, among other tasks, as indicated in the recently published Volume 1 of *Guidelines for Smart Grid Cyber Security* by the National Institute of Standards and Technology (NIST).[89] The complexity of such an endeavor, coupled with the amalgam of technologies and standards that will coexist in the development of the smart grid, makes it extremely necessary to have a common platform of development with flexibility and reliable performance.

FPGA DPs share these advantages, not to mention the fact that a single silicon FPGA chip can be used to study several smart grid technologies and their implementations. FPGA chips offer significant potential for application in the smart grid for performing encryption and decryption, intrusion detection, low-latency routing, data acquisition and signal processing, parallelism, configurability of hardware devices, and high-performance and high-bandwidth tamper-resistant applications. Dr. William Sanders, a member of the Smart Grid Advisory Committee of the NIST, has been among the most influential recently in the research on smart grid security. His research team and several collaborating universities proposed the use of a Trustworthy Cyber Infrastructure for the Power Grid (TCIPG) that focuses on the security of low-level devices and communications, as well as trustworthy operation of the power grid under a variety of conditions, including cyberattacks and emergencies.[92] TCIPG proposes a coordinated response and detection at multiple layers of the cyber infrastructure hierarchy, including but not limited to sensor/actuator and substation levels. At these levels of the hierarchy, SDR and wireless communications technologies could be used and studied to prevent attacks such as wireless jamming. Sanders et al. also proposed the use of specification-based IDS in protecting advanced metering infrastructures (AMIs).[93] A distributed FPGA-based network with adaptive and cooperative capabilities can be used to study several security and communication aspects of this infrastructure from the point of view of both the attackers and the defenders.

5.6 Conclusions

In this chapter, an innovative approach of employing a cognitive radio network for efficient management of information flow in the smart grid was presented. An outline of cognitive radio and the recently established IEEE 802.22 standard for WRANs was given. Existing and new hardware platforms for the innovative network test bed being built at Tennessee Technological University were described. To efficiently process the high-dimensional data in cognitive radio networks, dimensionality reduction techniques such as PCA, KPCA, and LMVU can be used. The SVM method was applied to a spectrum-monitoring example in Wi-Fi networks, and it was shown that better performance is achieved using dimensionality reduction for preprocessing the data. The recently developed robust PCA algorithm was presented for recovering a low-rank matrix when it was grossly corrupted with a sparse matrix of arbitrarily large magnitude. For the blind recovery of smart meter wireless transmissions in the presence of strong wideband interference, the robust PCA was used as a preprocessing method before applying an ICA-based algorithm. Finally, the vital issue of security in the smart grid was discussed, along with a possible approach to achieve this by employing FPGA-based fuzzy logic intrusion detection.

References

1. J. Mitola III and G. Maguire Jr., Cognitive radio: making software radios more personal, *IEEE Personal Communications* 6(4), 13–18 (1999).
2. S. Haykin, Cognitive radio: brain-empowered wireless communications, *IEEE Journal on Selected Areas in Communications* 23(2), 201–220 (2005).

3. G. Ganesan, Y. Li, B. Bing, and S. Li, Spatiotemporal sensing in cognitive radio networks, *IEEE Journal on Selected Areas in Communications* 26 (1), 5–12 (2008).

4. J. Bazerque and G. Giannakis, Distributed spectrum sensing for cognitive radio networks by exploiting sparsity, *IEEE Transactions on Signal Processing* 58(3), 1847–1862 (2010).

5. C. Cordeiro, K. Challapali, D. Birru, S. Shankar, et al., IEEE 802.22: an introduction to the first wireless standard based on cognitive radios, *Journal of Communications* 1(1), 38–47 (2006).

6. C. Cordeiro, K. Challapali, D. Birru, S. Shankar, et al. IEEE 802.22: the first worldwide wireless standard based on cognitive radios. In *2005 First IEEE International Symposium on New Frontiers in Dynamic Spectrum Access Networks, 2005. DySPAN 2005,* pp. 328–337. IEEE, New York (2005).

7. C. Cordeiro, K. Challapali, and M. Ghosh. Cognitive PHY and MAC layers for dynamic spectrum access and sharing of TV bands. In *Proceedings of the First International Workshop on Technology and Policy for Accessing Spectrum,* p. 3. ACM, New York (2006).

8. C. Stevenson, G. Chouinard, Z. Lei, W. Hu, S. Shellhammer, and W. Caldwell, IEEE 802.22: the first cognitive radio wireless regional area network standard, *IEEE Communications Magazine* 47(1), 130–138 (2009).

9. R. Qiu, A Cognitive Radio Network Testbed. Office of Naval Research (ONR) DURIP. N00010-10-0810. 2010.

10. R. C. Qiu, Smart Grid Research at TTU. Presented at Argonne National Laboratory (February 2010). Available at http://iweb.tntech.edu/rqiu/publications.htm

11. R. C. Qiu, Cognitive Radio and Smart Grid. Presented at IEEE Chapter, Huntsville, AL. (February 18, 2010). Available at http://iweb.tntech.edu/rqiu/publications.htm

12. R. C. Qiu, Z. Chen, N. Guo, Y. Song, P. Zhang, H. Li, and L. Lai, Towards a real-time cognitive radio network testbed: architecture, hardware platform, and application to smart grid. Presented at *Proceedings of the Fifth IEEE Workshop on Networking Technologies for Software-Defined Radio and White Space,* Boston (June 2010).

13. D. Raychaudhuri, I. Seskar, M. Ott, S. Ganu, K. Ramachandran, H. Kremo, R. Siracusa, H. Liu, M, and Singh. Overview of the ORBIT radio grid testbed for evaluation of next-generation wireless network protocols. In *Proceedings of IEEE Wireless Communications and Networking Conference,* New Orleans, LA, March 13–17, 2005. pp. 1664–1669 (2005).

14. I. Broustis, J. Eriksson, S. Krishnamurthy, and M. Faloutsos. A blueprint for a manageable and affordable wireless testbed: design, pitfalls and lessons learned. In *Proceedings of 3rd International Conference on Testbeds and Research Infrastructure for the Development of Networks and Communities,* May 21–23, 2007. pp. 1–6 (2007).

15. T. R. Newman, S. S. Hasan, D. Depoy, T. Bose, and J. H. Reed, Designing and deploying a building-wide cognitive radio network testbed, *IEEE Communications Magazine* 48(9), 106–112 (2010).

16. Ettus Research LLC, Home page (July 2010). http://www.ettus.com/.
17. GNU Radio, Home page (July 2010). http://www.gnuradio.org/.
18. Z. Chen, N. Guo, and R. C. Qiu, Experimental validation of channel state prediction considering delays in practical cognitive radio, *IEEE Transactions on Vehicular Technology* 16(4), 1314–1325 (2011).
19. Lyrtech Incorporated, *Small Form Factor SDR Evaluation Module/Development Platform Users Guide.* Lyrtech, Quebec City, Canada (February 2010).
20. Lyrtech Incorporated, *ADACMaster III Users Guide.* Lyrtech, Quebec City, Canada (January 2009).
21. Lyrtech Incorporated, Home page (July 2010). http://www.lyrtech.com/.
22. K. Amiri, Y. Sun, P. Murphy, C. Hunter, J. Cavallaro, and A. Sabharwal, WARP, a unified wireless network testbed for education and research. In *IEEE International Conference on Microelectronic Systems Education*, San Diego, CA, June 3–4, 2007. pp. 53–54 (2007).
23. Rice University, Home page (July 2010). http://warp.rice.edu/.
24. Mango Communications, Home page (September 2010). http://www.mangocomm.com/.
25. K. Tan, J. Zhang, J. Fang, H. Liu, Y. Ye, S. Wang, Y. Zhang, H. Wu, W. Wang, and G. Voelker, Sora: high performance software radio using general purpose multi-core processors. In *Proceedings of the 6th USENIX symposium on Networked Systems Design and Implementation*, pp. 75–90. USENIX Association, Berkeley, CA (2009).
26. K. Chowdhury and T. Melodia, Platforms and testbeds for experimental evaluation of cognitive ad hoc networks, *IEEE Communications Magazine.* 48(9), 96–104 (2010).
27. Z. Chen, N. Guo, and R. C. Qiu, Building a cognitive radio network testbed, *Proceedings of IEEE Southeastcon.* Nashville, TN (March 2011).
28. R. C. Qiu, *Cognitive Radio Network Testbed.* Funded research proposal for Defense University Research Instrumentation Program (DURIP) (August 2009). http://www.defense.gov/news/Fiscal 2010 DURIP Winners List.pdf
29. R. C. Qiu, Cognitive Radio and Smart Grid. Invited presentation at IEEE Chapter (February 2010). http://iweb.tntech.edu/rqiu.
30. Robert C. Qiu (PI). Cognitive Radio Institute. Funded research proposal for 2010 Defense Earmark (2010). http://www.opensecrets.org/politicians/earmarks.php?cid=N00003126
31. R. Qiu, Z. Hu, G. Zheng, Z. Chen, and N. Guo. Cognitive radio network for the smart grid: experimental system architecture, control algorithms, security, and microgrid testbed, *IEEE Transactions on Smart Grid* 2(4), 724–740 (2011).
32. R. C. Qiu, M. C. Wicks, Z. Hu, L. Li, and S. J. Hou, Wireless tomography (1): a novel approach to remote sensing. In *5th International Waveform Diversity and Design Conference*, Niagara Falls, Canada (August 2010).
33. T. Guo and S. A. Jafar, Degrees of freedom of the K user M N MIMO interference channel, *IEEE Transactions on Information Theory* 56, 12 (2010).

34. V. R. Cadambe and S. A. Jafar, Interference alignment and spatial degrees of freedom for the k user interference channel. In *IEEE International Conference on Communications, 2008. ICC'08*, pp. 971–975, Beijing (May 2008).

35. M. A. Maddah-Ali, A. S. Motahari, and A. K. Khandani, Communication over MIMO X channels: Interference alignment, decomposition, and performance analysis, *IEEE Transactions on Information Theory* 54(8), 3457–3470 (2008).

36. B. Nazer, S. A. Jafar, M. Gastpar, and S. Vishwanath, Ergodic interference alignment. In *IEEE International Symposium on Information Theory, 2009. ISIT 2009*, pp. 1769–1773, Seoul, Korea (2009).

37. C. Huang and S. A. Jafar, Degrees of freedom of the MIMO interference channel with cooperation and cognition, *IEEE Transactions on Information Theory* 55(9), 4211–4220 (2009).

38. C. S. Vaze and V. M. K. The degrees of freedom region of the MIMO cognitive interference channel with no CSIT. In *ISIT*, pp. 440–444, Austin, TX (June 2010).

39. J. B. Tenenbaum, V. Silva, and J. C. Langford, A global geometric framework for nonlinear dimensionality reduction, *Science* 290(5500), 2319–2323 (2000).

40. S. Roweis and L. Saul, Nonlinear dimensionality reduction by locally linear embedding, *Science* 290(5500), 2323–2326 (2000).

41. E. Keogh, K. Chakrabarti, M. Pazzani, and S. Mehrotra, Dimensionality reduction for fast similarity search in large time series databases, *Knowledge and Information Systems* 3(3), 263–286 (2001).

42. M. L. Raymer, W. F. Punch, E. D. Goodman, L. A. Kuhn, and A. K. Jain, Dimensionality reduction using genetic algorithms, *IEEE Transactions on Evolutionary Computation* 4(2), 164–171 (2002).

43. I. Jolliffe, *Principal Component Analysis*. Springer-Verlag, New York (2002).

44. B. Scholkopf, A. Smola, and K. Muller, Nonlinear component analysis as a kernel eigenvalue problem, *Neural Computation* 10(5), 1299–1319 (1998).

45. K. Weinberger and L. Saul, Unsupervised learning of image manifolds by semidefinite programming, *International Journal of Computer Vision* 70(1), 77–90 (2006).

46. K. Weinberger, B. Packer, and L. Saul, Nonlinear dimensionality reduction by semidefinite programming and kernel matrix factorization. In *Proceedings of the Tenth International Workshop on Artificial Intelligence and Statistics*, pp. 381–388, Barbados (January 2005).

47. G. Baudat and F. Anouar, Kernel-based methods and function approximation. In *International Joint Conference on Neural Networks, 2001. Proceedings. IJCNN'01*, vol. 2, pp. 1244–1249, Washington, DC (July 2001).

48. G. Wu, E. Y. Chang, and N. Panda, Formulating distance functions via the kernel trick. In *Proceedings of the Eleventh ACM SIGKDD International Conference on Knowledge Discovery in Data Mining*, pp. 703–709, Chicago (August 2005).

49. J. Mariéthoz and S. Bengio, A kernel trick for sequences applied to text-independent speaker verification systems, *Pattern Recognition* 40(8), 2315–2324 (2007).

50. J. Wang, J. Lee, and C. Zhang, Kernel trick embedded Gaussian mixture model. In *Algorithmic Learning Theory*, vol. 2842, pp. 159–174. SpringerLink, New York (2003).

51. L. Vandenberghe and S. Boyd, Semidefinite programming, *SIAM Review* 38(1), 49–95 (1996).

52. F. Alizadeh, J. P. A. Haeberly, and M. L. Overton, Primal-dual interior-point methods for semidefinite programming: convergence rates, stability and numerical results, *SIAM Journal on Optimization* 8(3), 746–768 (1998).

53. H. Wolkowicz, R. Saigal, and L. Vandenberghe, *Handbook of Semidefinite Programming: Theory, Algorithms, and Applications*. Springer-Verlag, Dordrecht (2000).

54. S. P. Boyd and L. Vandenberghe, *Convex Optimization*. Cambridge University Press (2004).

55. G. R. G. Lanckriet, N. Cristianini, P. Bartlett, L. E. Ghaoui, and M. I. Jordan, Learning the kernel matrix with semidefinite programming, *The Journal of Machine Learning Research* 5, 27–72 (2004).

56. S. J. Hou, Z. Qiu, R. Chen, and Z. Hu, Spectrum sensing using SVM and dimensionality reduction with experimental validation, http://arXiv.org/abs/1106.2325 (2011).

57. S. Haykin, D. Thomson, and J. Reed, Spectrum sensing for cognitive radio, *Proceedings of the IEEE* 97(5), 849–877 (May 2009). doi: 10.1109/JPROC.2009.2015711.

58. J. Ma, G. Y. Li, and B. H. Juang, Signal processing in cognitive radio, *Proceedings of the IEEE* 97(5), 805–823 (2009). doi: 10.1109/JPROC.2009.2015707.

59. D. Cabric, S. Mishra, and R. Brodersen. Implementation issues in spectrum sensing for cognitive radios. In *Proceedings of Conference Record of the Thirty-Eighth Asilomar Conference on Signals, Systems and Computers*, vol. 1, 772–776 (2004).

60. Z. Chen and R. C. Qiu, Prediction of channel state for cognitive radio using higher-order hidden Markov model. In *Proceedings of the IEEE Southeastcon*, pp. 276–282 (March 2010).

61. I. Jolliffe, *Principal Component Analysis*, 2nd edition. Springer-Verlag, New York (2002).

62. E. Candès, X. Li, Y. Ma, and J. Wright, Robust principal component analysis? *Journal of ACM (JACM)*, 58(3), 1–37 (May 2011).

63. J. Wright, A. Ganesh, S. Rao, and Y. Ma, Robust principal component analysis: Exact recovery of corrupted low-rank matrices via convex optimization, in *Proceedings of the Conference on Neural Information Processing Systems* (NIPS) (December 2009).

64. V. Chandrasekaran, S. Sanghavi, P. Parrilo, and A. Willsky, Rank-sparsity incoherence for matrix decomposition, *SIAM Journal on Optimization*, 21(2), 572–596 (2011).

65. E. Candès and T. Tao, The power of convex relaxation: near-optimal matrix completion, *IEEE Transactions on Information Theory* 56(5), 2053–2080 (2010).

66. E. Candès and B. Recht, Exact matrix completion via convex optimization, *Foundations of Computational Mathematics* 9(6), 717–772 (2009).

67. E. Candès and Y. Plan, Matrix completion with noise, *Proceedings of the IEEE* 98(6), 925–936 (2010).

68. B. Recht, M. Fazel, and P. Parrilo, Guaranteed minimum-rank solutions of linear matrix equations via nuclear norm minimization, *Arxiv preprint* arXiv:0706.4138 (2007).

69. B. Recht, W. Xu, and B. Hassibi. Necessary and sufficient conditions for success of the nuclear norm heuristic for rank minimization. In *47th IEEE Conference on Decision and Control, 2008. CDC 2008*, pp. 3065–3070. IEEE, New York (2009).

70. J. Cai, E. Candès, and Z. Shen, A singular value thresholding algorithm for matrix completion, *Arxiv preprint* arXiv:0810.3286 (2008).

71. Z. Lin, M. Chen, L. Wu, and Y. Ma, The augmented Lagrange multiplier method for exact recovery of corrupted low-rank matrices, UIUC Technical Report UILU-ENG-09-2215 (November 2009).

72. M. Chen. http://perception.csl.illinois.edu/matrix-rank/sample code.html.

73. S. Marple Jr., *Digital spectral analysis with applications*. Prentice Hall, Englewood Cliffs, NJ (1987).

74. P. Comon, Independent component analysis, a new concept? *Signal Processing* 36(3), 287–314 (1994).

75. A. Hyvarinen and E. Oja, One-unit learning rules for independent component analysis. In *Advances in Neural Information Processing Systems*, 480–486. Morgan Kaufmann, New York (1997).

76. A. Hyvarinen and E. Oja, Independent component analysis: algorithms and applications, *Neural Networks* 13(4–5), 411–430 (2000).

77. A. Hyvarinen, J. Karhunen, and E. Oja, *Independent Component Analysis*. Wiley, New York (2001).

78. H. Liao and D. Niebur, Load profile estimation in electric transmission networks using independent component analysis, *IEEE Transactions on Power Systems* 18(2), 707–715 (2003).

79. D. Pham, Blind separation of instantaneous mixture of sources via an independent component analysis, *IEEE Transactions on Signal Processing* 44(11), 2768–2779 (2002).

80. T. Lee, M. Lewicki, and T. Sejnowski, ICA mixture models for unsupervised classification of non-Gaussian classes and automatic context switching in blind signal separation, *IEEE Transactions on Pattern Analysis and Machine Intelligence* 22(10), 1078–1089 (2002).

81. S. Amari, A. Cichocki, and H. Yang. A new learning algorithm for blind signal separation. In *Advances in Neural Information Processing Systems*, 757–763. Morgan Kaufman, New York (1996).

82. Z. Koldovsky and P. Tichavsky, Blind instantaneous noisy mixture separation with best interference-plus-noise rejection. In *Proceedings of the 7th International Conference on Independent Component Analysis and Signal Separation*, pp. 730–737. Springer-Verlag, New York (2007).

83. L. Husheng, M. Rukun, L. Lifeng, and R. Qiu, Compressed meter reading for delay-sensitive and secure load report in smart grid. In *First IEEE International Conference on Smart Grid Communications, 2010. SmartGridComm 2010*, pp. 114–119. Gaithersburg, MD (October 2010).

84. D. Donoho, Compressed sensing, *IEEE Transactions on Information Theory* 52(4), 1289–1306 (2006).

85. E. Candès, J. Romberg, and T. Tao, Robust uncertainty principles: exact signal reconstruction from highly incomplete frequency information, *IEEE Transactions on Information Theory* 52(2), 489–509 (2006).

86. S. Chen, D. Donoho, and M. Saunders, Atomic decomposition by basis pursuit, *SIAM Review* 43(1), 129–159 (2001).

87. R. Ranganathan and W. B. Mikhael, A comparative study of complex gradient and fixed-point ICA algorithms for interference suppression in static and dynamic channels, *Signal Processing* 88(2), 399–406 (2008). doi: 10.1016/j.sigpro.2007.08.002. http://www.sciencedirect.com/science/article/B6V18-4PF1W9K-2/2/f6fede5f cdf79d0b75c0b5d050020861.

88. National Energy Technology Laboratory, *A Systems View of the Modern Grid*. Department of Energy, Washington, DC (January 2007).

89. National Institute of Standards and Technology, *Guidelines for Smart Grid Security*: vol. 1, *Smart Grid Cyber Security Strategy, Architecture, and High-Level Requirements*, The Smart Grid Interoperability Panel-Cyber Security Working Group, August 2010.

90. M. Pazos-Revilla and A. Siraj, An experimental model of an FPGA-based intrusion detection systems. In *2011 International Conference on Computers and Their Applications*, New Orleans, LA (March 2011).

91. M. Pazos-Revilla, FPGA Based Fuzzy Intrusion Detection System for Network Security, master's thesis, Tennessee Technological University, Cookeville (2010).

92. W. Sanders, *TCIP: Trustworthy Cyber Infrastructure for the Power Grid*, Technical report. Information Trust Institute, University of Illinois at Urbana-Champaign (2011).

93. R. Berthier, W. Sanders, and H. Khurana, Intrusion detection for advanced metering infrastructures: requirements and architectural directions. In *First IEEE International Conference on Smart Grid Communications (SmartGridComm), 2010*, pp. 350–355. IEEE, New York (2010).

PART 2

SECURITY AND PRIVACY IN SMART GRIDS

6

REQUIREMENTS AND CHALLENGES OF CYBERSECURITY FOR SMART GRID COMMUNICATION INFRASTRUCTURES

ROSE QINGYANG HU AND YI QIAN

Contents

Upgrading an existing power grid into a smart grid requires significant dependence on intelligent and secure communication infrastructures. It requires systematic security frameworks for distributed communications, pervasive computing, and sensing technologies in the smart grid. However, as many of the communication technologies currently recommended for use by a smart grid are vulnerable to cyberattacks, it could lead to unreliable system operations, causing unnecessary expenditures, even consequential disaster for both utilities and consumers. In this chapter, we summarize the possible vulnerabilities and the cyber-security requirements in smart grid communications and discuss the challenges of cyber security for smart grid communications.

6.1 Introduction

A smart grid communication system is comprised of several subsystems. It is eventually a network of networks. A supervisory control and data acquisition (SCADA) system is not only a controlling system but also a communication network in a smart grid. The communication networks in smart grid systems could include dedicated or overlay land mobile radios (LMRs), cellular, microwave, fiber-optic, wired lines such as power line communication (PLC), RS-232/RS-485 serial links, wireless local-area networks (WLANs) media or a versatile data network combining these media. In this section, we briefly discuss the background of a smart grid system in several aspects: SCADA system, communication networks, deployments of secure smart grid communications, and high-level security requirements. Figure 6.1 shows a typical smart grid communication system.[1]

6.1.1 Background

Core to the monitoring and control of a substation is the SCADA system. It is utilized for distribution automation (DA) and computerized remote control of medium-voltage (MV) substations and power grids, and it helps electric utilities achieve higher supply reliability and reduces operating and maintenance costs. In the past, sectionalizer switchgears, ring main units, reclosers, and capacitor banks were designed for local operations with limited remote control. Today, using

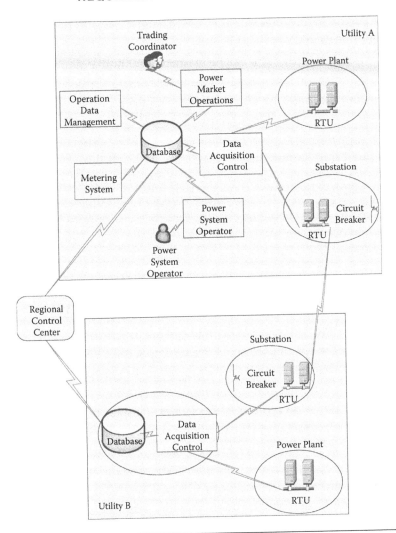

Figure 6.1 A typical smart grid communication system. (From C. H. Hauser, D. E. Bakken, and A. Bose, *IEEE Power and Energy Magazine*, pp. 47–55, March–April 2005. With permission.)

SCADA over reliable wireless communication links, remote terminal units (RTUs) provide powerful integrated solutions when upgrading remotely installed electric equipment. In a distribution management system (DMS), RTUs seamlessly interface via SCADA with a wide range of high-performance control centers supplied by leading vendors worldwide. Connection to these energy management systems (EMSs) and DA/DMS control centers is typically provided via a high-performance Internet Protocol (IP) gateway or a similar node.[2]

Different scales and structures of smart grid systems adopt different communication networking solutions. Advanced metering infrastructure (AMI) solutions can be meshed or point to point, with short local coverage or long-range communications.[3,4] Options for backhaul solutions might be fiber, wireless broadband, or broadband over a power line. The possible solutions include WiMAX, WLAN, wireless sensor network (WSN), cellular, and LMR, depending on the reliability, throughput, and coverage desired by the utility. The wireless communication solutions can be either licensed or unlicensed, again depending on the needs of the utility. For the highest reliability, licensed solutions should be chosen. Each of these options has advantages and disadvantages, but what is consistently true of any and all of the solutions is the need to have a scalable security solution.[5]

Smart grid deployments must meet stringent security requirements. Strong authentication will be required for all users and devices that may affect the operation of the grid. With the large number of users and devices affected, scalable key and trust management systems, customized to the specific needs of the energy service provider, will be essential. What has been learned from years of deploying and operating large secure network communication systems is that the effort required to provision symmetric keys into thousands of devices can be too expensive or insecure. The development of key and trust management systems for large networks is required; these systems can be leveraged from other industries, such as LMR systems and Association of Public-Safety Communications Officials (APCO) radio systems. Several APCO-deployed systems provide statewide wireless coverage, with tens of thousands of secure devices. Trust management systems, based on public key infrastructure (PKI) technology, could be customized specifically for smart grid operators, easing the burden of providing security that adheres to the standards and guidelines that are known to be secure.[6]

6.1.2 High-Level Requirements

According to the Electric Power Research Institute (EPRI), one of the biggest challenges facing smart grid deployment is related to cybersecurity of the systems.[7] According to the EPRI report, cybersecurity is a critical issue due to the increasing potential of cyberattacks and incidents against this critical sector as it becomes increasingly intercon-nected. Cybersecurity must address not only deliberate attacks, such as from disgruntled employees, industrial espionage, or terrorists, but also inadvertent compromises of the information infrastructure due to user errors, equipment failures, and natural disasters. Vulnerabilities might allow an attacker to penetrate a network, gain access to control software, and alter load conditions to destabilize the grid in unpredict-able ways. The high-level requirements for smart grid communication security are conducted in various organizations and the correspond-ing standards in details. The cybersecurity requirements for smart grid communications are discussed further in the rest of this chapter.

There are many organizations working on the development of smart grid security requirements, including the North American Electrical Reliability Corporation Critical Infrastructure Protection (NERC CIP), International Society of Automation (ISA), Institute of Electrical and Electronics Engineers (IEEE) (IEEE 1402), National Infrastructure Protection Plan (NIPP), and National Institute of Standards and Technology (NIST), which has a number of smart grid cybersecurity programs ongoing.

One prominent source of requirements is the Smart Grid Interoperability Panel (SGiP) Cyber Security Working Group, previously the NIST Cyber Security Coordination Task Group (CSCTG).[8] The NIST CSCTG was established to ensure consistency in the cybersecurity requirements across all the smart grid domains and components. The latest draft document from the Cyber Security Working Group, NIST Interagency Report (NIST-IR7628),[9] *Smart Grid Cyber Security Strategy and Requirements*, continues to evolve at the time of this writing. NIST and the Department of Energy (DoE) GridWise Architecture Council (GWAC)[10] have established domain expert working groups (DEWGs): Home-to-Grid (H2G), Building-to-Grid (B2G), Industrial-to-Grid (I2G), Transmission and Distribution (T&D), and Business and Policy (B&P).

Working with standards bodies, such as NIST and others, will be extremely important to ensure a highly secure, scalable, consistently deployed smart grid system as these standards bodies will drive the security requirements of the system.[11]

One thing is consistent among the various standards bodies: The security of the grid will strongly depend on authentication, authorization, and privacy technologies. Privacy technologies are well matured. The Advanced Encryption Standard (AES)[12] and Triple Data Encryption Algorithm (3DES)[13] solutions approved by the Federal Information Processing Standard (FIPS), offering strong security and high performance, are readily available. The specific privacy solution required will depend on the type of communication resource protected. As a specific example, NIST has determined that the 3DES solution will likely become insecure by the year 2030. Considering that utility components are expected to have long lifetimes, the AES would be the preferred solution for new components. However, it is reasonable to expect that under certain circumstances when legacy functionality must be supported and the risk of compromise is acceptable, 3DES could be used.

Wireless links will be secured with technologies from well-known standards such as IEEE 802.11i[14] and IEEE 802.16e.[15] Different wireless protocols have varying degrees of security mechanisms. Wired links will be secured with firewalls, virtual private networks (VPNs), and IPSec (Internet Protocol Security) technologies. Higher-layer security mechanisms such as Secure Shell (SSH) and Secure Sockets Layer/Transport Layer Security (SSL/TLS) should also be used.[16]

System architects and designers often identify the need for and specify the use of secure protocols, such as SSH and IPSec, but then skip the implementation details associated with establishing security associations between end points of communications. Such an approach is likely to result in a system in which the necessary procedures for secure key management can quickly become an operational nightmare. This is because, when system architects do not develop an integrated and comprehensive key management scheme, customers may be provided with few key management options and often resort to manually preconfiguring symmetric keys. This approach is simple for the system designers, but it can be very expensive for the system owners/operators.

6.2 Vulnerabilities and Security Requirements

The reliability of a smart grid depends on the reliability of the control and communication systems. For development of smart grid systems, the communication systems are becoming more sophisticated, allowing for better control and higher reliability. The smart grid will require

Table 6.1   Layered Security Protocols

LAYER	SECURITY PROTOCOL	APPLICATION	CONFIDENTIALITY	INTEGRITY	AUTHENTICATION
Application	WS-Security	Document	Yes	Yes	Data
	PGP/GnuPG	E-mail	Yes	Yes	Message
	S/MIME		Yes	Yes	
	HTTP digest authentication	Client to service	No	No	User
Transport	SSH		Yes	Yes	Server
	SSL/TLS		Yes	Yes	
Network	IPSec	Host to host	Yes	Yes	Host
Link	CHAP/PAP	Point to point	No	No	Client
	WEP/WAP/802.1X	Wireless access	Yes	Yes	Device

Source: From Y. Yan, Y. Qian, H. Sharif, and D. Tipper, *IEEE Communications Surveys and Tutorials,* vol. 14(4), pp. 998–1010, 2012. With permission from IEEE.[17]

Note: CHAP/PAP = Challenge Handshake Authentication Protocol/Password Authentication Protocol, HTTP = Hypertext Transfer Protocol, PGP/GnuPG = pretty good privacy/Gnu Privacy Guard, S/MIME = secure/multipurpose Internet mail extensions, WEP/WAP = wired equivalent privacy/WiFi protected access, WS-Security = web services security.

higher degrees of network connectivity to support the new features. The higher degree of connectivity should have sophisticated security protocols to deal with the vulnerabilities and security breaches. Table 6.1 lists some security protocols adopted by different layers in communication networks with the specific security requirements; more details were summarized by Dzung et al.[18] In this section, we discuss the major security vulnerabilities and requirements in privacy, availability, integrity, authentication, authorization, auditability, non-repudiability, third-party protection, and trust components for smart grid communication security.

6.2.1 Privacy

Privacy issues have to be covered with the derived customer consumption data as they are created in metering devices. Consumption data contain detailed information that can be used to gain insights on a customer's behavior. Smart grid communications have unintended consequences for customer privacy. Electricity usage information stored at the smart meter and distributed thereafter acts as an information-rich side channel, exposing customers' habits and behaviors. Certain activities, such as watching television, have detectable power consumption signatures. History has shown that where financial or political incentives align, the techniques for mining behavioral data will evolve quickly to match the desires of those who would exploit that information.[19]

Utility companies are not the only sources of potential privacy abuse. The recently announced Google PowerMeter service,[20] for instance, receives real-time usage statistics from installed smart meters. Customers subscribing to the service receive a customized Web page that visualizes local usage. Although Google has yet to announce the final privacy policy for this service, early versions leave the door open to the company to use this information for commercial purposes, such as marketing individual or aggregate usage statistics to third parties. Although services such as Google PowerMeter are optional, customers have less control over the use of power information delivered to utility companies. Existing privacy laws in the United States are in general a patchwork of regulations and guidelines. It is unclear how these or any laws apply to customer energy usage yet.

6.2.2 Availability

Availability refers to ensuring that unauthorized persons or systems cannot deny access or use to authorized users. For smart grid systems, this refers to all the information technology (IT) elements of the plant, like control systems, safety systems, operator workstations, engineering workstations, manufacturing execution systems, as well as the communication systems between these elements and to the outside world.

Malicious attacks targeting availability can be considered as denial-of-service (DoS) attacks, which attempt to delay, block, or even corrupt information transmission to make network resources unavailable to communicating nodes that need information exchange in the smart grid. Since it is widely expected that at least part, if not all, of the smart grid will use IP-based protocols (e.g., International Electrotechnical Commission [IEC] 61580 has already adopted the Transmission Control Protocol [TCP]/IP as a part of its protocol stacks[21]), and TCP/IP is vulnerable to DoS attacks. DoS attacks against TCP/IP have been well studied in the literature regarding attacking types, prevention, and response.[22–24]

However, a major difference between a smart grid communication network and the Internet is that the smart grid is more concerned with the message delay than the data throughput due to the timing constraint of messages transmitted over the power networks. Indeed, network traffic in smart grid communication networks is in general time critical. For instance, the delay constraint of Generic Object Oriented Substation Event (GOOSE) messages is 4 ms in IEC 61850.[9]

Intruders only need to connect to communication channels rather than authenticated networks in the smart grid; it is very easy for them to launch DoS attacks against the smart grid communication networks, especially for the wireless-based communication networks that are susceptible to jamming attacks.[25–27] Hence, it is of critical importance to evaluate the impact of DoS attacks on the smart grid and to design effective countermeasures to such attacks.

6.2.3 Integrity

Integrity refers to preventing undetected modification of information by unauthorized persons or systems. For smart grid communication systems, this applies to information such as product recipes, sensor values, or control commands. This objective includes defense against information modification via message injection, message replay, and message delay on the network. Violation of integrity may cause safety issues; that is, equipment or people may be harmed.

Differing from attacks targeting availability, attacks targeting data integrity can be regarded as less brute force and more sophisticated attacks. The target of the integrity attacks is either customer information (e.g., pricing information and customer account balance) or network operation information (e.g., voltage readings, device running status). In other words, such attacks attempt to deliberately modify the original information in the smart grid communication system to corrupt critical data exchange in the smart grid.

The risk of attacks targeting data integrity in the power networks is indeed real. A notable example is the recent work by Liu et al.,[28] which proposed a new type of attacks, called false data injection attacks, against the state estimation in the power grid. It assumed that an attacker has already compromised one or several meters and pointed out that the attacker can take advantage of the configuration of a power system to launch attacks by injecting false data to the monitoring center, which can legitimately pass the data integrity check used in current power systems.

6.2.4 Authentication

Authentication is concerned with determination of the true identity of a communication system participator and mapping of this identity to a system-internal principal (e.g., valid user account) by which this user is known to the system. Most other security objectives, most notably authorization, distinguish between legitimate and illegitimate users based on authentication.

6.2.5 Authorization

Authorization, also known as access control, is concerned with preventing access to the system by persons or systems without permission to do so. In the wider sense, authorization refers to the mechanism that distinguishes between legitimate and illegitimate users for all other security objectives (e.g., confidentiality, integrity, etc.). In the narrower sense of access control, it refers to restricting the ability to issue commands to the plant control system. Violation of authorization may cause safety issues.

6.2.6 Auditability

Auditability is concerned with being able to reconstruct the complete history of the system behavior from historical records of all (relevant) actions executed on it. This security objective is mostly relevant to discover and find reasons for malfunctions in the system after the fact and to establish the scope of the malfunction or the consequences of a security incident. Note that auditability without authentication may serve diagnostic purposes but does not provide accountability.

6.2.7 Nonrepudiability

Nonrepudiability refers to being able to provide irrefutable proof to a third party regarding who initiated a certain action in the system, even if this actor is not cooperating. This security objective is relevant to establish accountability and liability. In the context of smart grid systems, this is most important in reference to regulatory requirements. Violation of this security requirement typically has legal/commercial consequences.

6.2.8 Third-Party Protection

Third-party protection refers to averting damage done to third parties via the communication systems, that is, damage that does not involve safety hazards of the controlled plant itself. The successfully attacked and subverted automation system could be used for various attacks on the communication systems or data or users of external third parties (e.g., via distributed DoS [DDoS]) or worm attacks. Consequences could reach from a damaged reputation of a smart grid system owner to legal liability for the damages of the third party. The risk to third parties through possible safety-relevant failures of the plant arising out of attacks against the plant automation system is covered by other security objectives, most notably authorization/access control.

6.2.9 Trust

The new designs of future smart grid communication systems form a multilayer architecture. The growth of smart grid systems resulted in a plentifulness of power system-related software applications, developed in many different programming languages and platforms. Extending old applications or developing new ones usually involves integrating legacy systems. Therefore, approaching the security of future smart grid communication networks cannot be done with a completely new start.

In parallel to the development of smart grid communication systems, the complete and monolithic cybersecurity infrastructure is not a viable option. Instead, multilayer architecture, advanced control methodologies, and dependable software infrastructure as well as device protection mechanisms and hardware-monitoring anchors have to be specified at the same time. Advanced control approaches have to include predictive and self-adaptive intelligence at higher-level and cross-layer mapping to the different technical layers. The dependable software infrastructures have to be designed to identify and isolate higher-layer independent applications as well as to secure cross-layer communications. With such architecture, it should have the flexibility of incorporating parts of existing infrastructure with the frontiers and interfaces to adjacent systems. Furthermore, the architecture needs the flexibility to interchange or update the part of the system in a secure way at a later stage due to new laws and regulations or new developments in the energy market.[29]

6.3 Cybersecurity Challenges

There are many cybersecurity challenges for a secure smart grid communication system. The major challenges in building and operating a secure smart grid communication system include internetworking, security policy and operations, security services, and others.

6.3.1 Internetworking

The interconnected smart grid communication systems are riddled with vulnerabilities that vary across the networks due to the lack of built-in security in many applications and devices. This should not be the model for a network as important as the smart grid. Layers of defense should be built into the solution to minimize the threats from interruption, interception, modification, and fabrication.

Keeping the network private (i.e., with all transport facilities wholly owned by a utility) would greatly minimize the threats from intruders as there would be no potential for access from intruders over the Internet. But, having a completely separate network is not feasible in today's highly connected world. It makes good business sense to reuse communication facilities, such as the Internet. A minimally secured smart grid connected to the Internet, as commonly found with commercial networks, opens the grid to threats from multiple types of attacks. These include cyberattacks from hostile groups looking to cause an interruption to the power supply.[19,30]

One of these cyberattacks is worm infestations, which have proven to negatively impact critical network infrastructures. Such threats have largely been the result of leaving a network vulnerable to threats from the Internet. For example, there have been DoS attacks on a single network that disrupted all directory name servers, thus prohibiting users from connecting to any of the resources. It demonstrates the fragility of an interconnected smart grid communication infrastructure.[31]

All connections to the Internet from a smart grid network need to be highly secure. Intrusion detection is needed not only at the points where a smart grid network connects to the Internet but also at critical points within the network as well as vulnerable wireless interfaces.[32]

The components, systems, networks, and architecture are all important to the security design and reliability of the smart grid communication solutions. But, it is inevitable that an incident will occur at some point, and one must be prepared with the proper incident response plan. This can vary between commercial providers and private utility networks. A private utility network is likely to provide better consistency of the incident response plan in the event of a security incident, assuming the private network is built on a standardized framework of hardware and software. The speed of the response decreases exponentially as the number of parties involved increases. Conversely, a private network would ideally depend on fewer parties; therefore, a more efficient incident response process would provide for more rapid response and resolution. The rapidity of the response is critical during situations that involve a blackout.[33]

Criticalness of a device or a system also determines how prone it will be to attacks. History has shown that private networks by their inherent nature are less prone to attacks. As a result, it is recommended as the best approach when security is paramount.[34]

6.3.2 Security Policy and Operations

The reliability of a smart grid depends on the proper operations of many components and the proper connectivity between them.[35] To disrupt a smart grid system, an attacker might attempt to gain electronic access to a component and misconfigure it or to impersonate another component and report a false condition or alarm. One of the simplest type of attacks that an adversary might attempt is the DoS attack: The adversary prevents authorized devices from communicating by consuming excessive resources on one device. For example, it is a well-known issue that if a node, such as a server or an access control device, uses an authentication protocol that is prior to authentication and authorization, then the node may be subject to DoS attacks. Smart grid protocol designers must ensure that proper care and attention are given to this threat during protocol development.

Many organizations will be involved in the operations of a smart grid. As additional distributed intelligence is added to the network, it will be essential that entities (people or devices) can authenticate and determine the authorization status of other entities from a remote organization. This issue is commonly referred to as federated identity management. There are many possible technical solutions to this issue, such as those offered by Security Assertion Markup Language (SAML),[36] Web Services Trust (WS-Trust),[37] and PKI.[38] Not only will vendors need to offer consistent technical solutions, but also organizations will further need consistent security policies. Great care must be taken by organizations to ensure their security policies and practices are not in conflict with those of other organizations with which they will need interoperability. At least a minimum set of operational security policies for the organizations operating a smart grid is formally adopted and documented in industry standards.[39]

6.3.3 Security Services

Managing and maintaining a secure smart grid will be equally as vital as developing, deploying, and integrating a secure smart grid solution. Security services will help network operators identify, control, and manage security risks in smart grid communications. According to EPRI, every aspect of a smart grid must be secure.[6] Cybersecurity technologies are not enough to achieve secure operations without policies, ongoing risk assessment, and training. The development of these human-focused procedures takes time and needs to take time to ensure that they are done correctly. A smart grid requires access to cost-effective, high-performance security services, including expertise in mobility, security, and system integration. These security services can be tailored per utility to best fit their needs and help them achieve

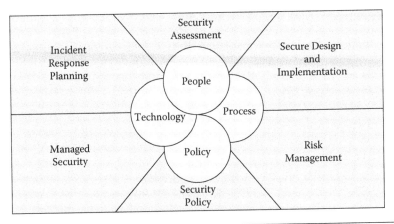

Figure 6.2   Smart grid security services. (From A. R. Metke and R. L. Ekl, in *Innovative Smart Grid Technologies (ISGT2010)*, pp. 1–7, Gaithersburg, MD, March 2010. With permission from IEEE.)

their organizational objectives. Figure 6.2 illustrates a typical set of security services in smart grid communications.[40]

6.4 Conclusions

In this chapter, we discussed the background and requirements as well as challenges for smart grid communication security. As a critical infrastructure, the smart grid requires comprehensive solutions for cybersecurity. A comprehensive communication architecture with security built in from the very beginning is necessary. A smart grid communication security solution requires a holistic approach, including traditional schemes such as PKI technology, trusted computing elements, and authentication mechanisms based on industry standards. Clearly, securing the smart grid communication infrastructure will require the use of standards-based state-of-the-art security protocols. To achieve the vision put forth, there are many steps that need to be taken. Primary among them is the need for a cohesive set of requirements and standards for smart grid security. Industry and other participants should continue the work that has begun under the direction of NIST to accomplish these foundational steps quickly. However, the proper attention must be paid to creating the requirements and standards as they will be utilized for many years, given the life cycle of utility components.

References

1. C. H. Hauser, D. E. Bakken, and A. Bose, A failure to communicate—next generation communication requirements, technologies, and architecture for electrical power grid, *IEEE Power and Energy Magazine*, pp. 47–55, March–April 2005.
2. S. Hong and M. Lee, Challenges and direction toward secure communication in the SCADA system, in *2010 Eighth Annual Communication Networks and Services Research Conference (CNSR)*, pp. 381–386, Montreal, May 2010.
3. L. Wenpeng, D. Sharp, and S. Lancashire, Smart grid communication network capacity planning for power utilities, in *2010 IEEE PES Transmission and Distribution Conference and Exposition*, pp. 1–4, New Orleans, LA, April 2010.
4. E. Liu, M. L. Chan, C. W. Huang, N. C. Wang, and C. N. Lu, Electricity grid operation and planning related benefits of advanced metering infrastructure, in *CRIS2010*, pp. 1–5, 2010.
5. P. P. Parikh, M. G. Kanabar, and T. S. Sidhu, Opportunities and challenges of wireless communication technologies for smart grid applications, in *IEEE Power and Energy Society General Meeting*, pp. 1–7, 2010.
6. A. R. Metke and R. L. Ekl, Security technology for smart grid networks, *IEEE Transactions on Smart Grid*, vol. 1, pp. 99–107, 2010.
7. National Institute of Standards and Technology, *Report to NIST on Smart Grid Interoperability Standards Roadmap EPRI*, June 17, 2009. Available at http://www.nist.gov/smartgrid/InterimSmartGridRoadmapNISTRestructure.pdf
8. Z. Tao, L. Weimin, W. Yufei, D. Song, S. Congcong, and C. Lu, The design of information security protection framework to support smart grid, *POWERCON 2010*, pp. 1–5, 2010.
9. National Institute of Standards and Technology, *Draft Smart Grid Cyber Security Strategy and Requirements, NIST IR 7628*, September 2009. Available at http://csrc.nist.gov/publications/drafts/nistir-7628/draft-nistir-7628.pdf
10. S. Widergren, A. Levinson, J. Mater, and R. Drummond, Smart grid interoperability maturity model, in *2010 IEEE Power and Energy Society General Meeting*, pp. 1–6, 2010.
11. S. Rohjans, M. Uslar, R. Bleiker, J. Gonzalez, M. Specht, T. Suding, and T. Weidelt, Survey of smart grid standardization studies and recommendations, in *IEEE SmartGridComm 2010*, pp. 583–588, 2010.
12. National Institute of Standards and Technology, *Announcing the Advanced Encryption Standard (AES)*, in Federal Information Processing Standards Publication 197, NIST, Gaithersburg, MD, November 26, 2001.
13. National Institute of Standards and Technology, *Data Encryption Standard*, Federal Information Processing Standards (FIPS) Publication 46-7, NIST, Gaithersburg, MD, 1999.

14. Institute of Electrical and Electronics Engineers, *IEEE Standard 802.11i, IEEE Standard for Information Technology-Telecommunications and Information Exchange Between Systems—Local and Metropolitan Area Networks—Specific Requirements Part 11: Wireless LAN Medium Access Control (MAC) and Physical Layer (PHY) Specifications Amendment 6: Medium Access Control (MAC) Security Enhancements*, pp. 1–175, IEEE, New York, 2004.

15. Institute of Electrical and Electronics Engineers, *IEEE Standard 802.16e, IEEE Standard for Local and Metropolitan Area Networks Part 16: Air Interface for Fixed and Mobile Broadband Wireless Access Systems Amendment 2: Physical and Medium Access Control Layers for Combined Fixed and Mobile Operation in Licensed Bands and Corrigendum 1*, pp. 1–822, IEEE, New York, 2006.

16. A. Bendahmane, M. Essaaidi, A. El Moussaoui, and A. Younes, Grid computing security mechanisms: State-of-the-art, in *International Conference on Multimedia Computing and Systems (ICMCS '09)*, pp. 535–540, 2009.

17. Y. Yan, Y. Qian, H. Sharif, and D. Tipper, *IEEE Communications Surveys and Tutorials*, vol. 14 (4), pp. 998–1010, 2012.

18. D. Dzung, M. Naedele, T. P. Von Hoff, and M. Crevatin, Security for Industrial Communication Systems, Proceedings of the IEEE, vol. 93, pp. 1152–1177, 2005.

19. P. McDaniel and S. McLaughlin, Security and privacy challenges in the smart grid, *IEEE Security and Privacy*, vol. 7, pp. 75–77, 2009.

20. K. Allan, Power to the people [power energy saving], *Engineering and Technology*, vol. 4, pp. 46–49, 2009.

21. T. S. Sidhu and Y. Yin, Modelling and simulation for performance evaluation of IEC61850-based substation communication systems, *IEEE Transactions on Power Delivery*, vol. 22, no. 3, pp. 1482–1489, July 2007.

22. C. L. Schuba, I. V. Krsul, M. G. Kuhn, E. H. Spafford, A. Sundaram, and D. Zamboni, Analysis of a denial of service attack on TCP, in *Proceedings of IEEE Symposium on Security and Privacy (S&P 1997)*, May 1997.

23. A. Yaar, A. Perrig, and D. Song, Pi: A path identification mechanism to defend against DDoS attacks, in *Proceedings of IEEE Symposium on Security and Privacy (S&P 2003)*, 2003.

24. J. Mirkovic and P. Reiher, A taxonomy of DDoS attack and DDoS defense mechanisms, *SIGCOMM Computer Communications Review*, vol. 34, no. 2, pp. 39–53, 2004.

25. M. Strasser, S. Capkun, C. Popper, and M. Cagalj, Jamming-resistant key establishment using uncoordinated frequency hopping, in *Proceedings of IEEE Symposium on Security and Privacy (S&P 2008)*, pp. 64–78, May 2008.

26. C. Popper, M. Strasser, and S. Capkun, Jamming-resistant broadcast communication without shared keys, in *Proceedings of the 18th USENIX Security Symposium (Security 09)*, August 2009.

27. Y. Liu, P. Ning, H. Dai, and A. Liu, Randomized differential DSSS: Jamming-resistant wireless broadcast communication, in *Proceedings of the 29th IEEE Conference on Computer Communications (INFOCOM 10)*, March 2010.

28. Y. Liu, P. Ning, and M. Reiter, False data injection attacks against state estimation in electric power grids, in *Proceedings of ACM Conference on Computer and Communications Security (CCS 09)*, September 2009.

29. N. Kuntze, C. Rudolph, M. Cupelli, J. Liu, and A. Monti, Trust infrastructures for future energy networks, in *IEEE Power and Energy Society General Meeting 2010*, pp. 1–7, 2010.

30. L. Husheng, M. Rukun, L. Lifeng, and R. C. Qiu, Compressed meter reading for delay-sensitive and secure load report in smart grid, in *IEEE SmartGridComm 2010*, pp. 114–119, 2010.

31. G. Carl, G. Kesidis, R. R. Brooks, and R. Suresh, Denial-of-service attack-detection techniques, *IEEE Internet Computing*, vol. 10, pp. 82–89, 2006.

32. S. Kent, On the trail of intrusions into information systems, *IEEE Spectrum*, vol. 37, pp. 52–56, 2000.

33. C. W. Ten, G. Manimaran, and C. C. Liu, Cybersecurity for critical infrastructures: attack and defense modeling, *IEEE Transactions on Systems, Man and Cybernetics, Part A: Systems and Humans*, vol. 40, no. 4, pp. 853–865, July 2010.

34. W. Dong, L. Yan, M. Jafari, P. Skare, and K. Rohde, An integrated security system of protecting smart grid against cyber attacks, in *Innovative Smart Grid Technologies (ISGT 2010)*, pp. 1–7, 2010.

35. M. Jensen, C. Sel, U. Franke, H. Holm, and L. Nordstrom, Availability of a SCADA/OMS/DMS system—A case study, in *IEEE Innovative Smart Grid Technologies Conference Europe (ISGT Europe 2010)*, pp. 1–8, 2010.

36. T. Komura, Y. Nagai, S. Hashimoto, M. Aoyagi, and K. Takahashi, Proposal of delegation using electronic certificates on single sign-on system with SAML-Protocol, in *Ninth Annual International Symposium on Applications and the Internet (SAINT '09)*, pp. 235–238, 2009.

37. C. Yongkai and T. Shaohua, Security scheme for cross-domain grid: integrating WS-Trust and grid security mechanism, in *International Conference on Computational Intelligence and Security(CIS '08)*, pp. 453–457, 2008.

38. R. Perlman, An overview of PKI trust models, *IEEE Network*, vol. 13, pp. 38–43, 1999.

39. R. J. Thomas, Putting an action plan in place, *IEEE Power and Energy Magazine*, vol. 7, pp. 26–31, 2009.

40. A. R. Metke and R. L. Ekl, Smart grid security technology, in *Innovative Smart Grid Technologies (ISGT2010)*, pp. 1–7, Gaithersburg, MD, March 2010.

7

REGULATIONS AND STANDARDS RELEVANT FOR SECURITY OF THE SMART GRID

STEFFEN FRIES AND HANS-JOACHIM HOF

Contents

Cyber attacks on critical infrastructures are increasingly becoming a threat to societies around the world. Hence, governments and standardization organizations are defining and improving their regulation and standards framework for one of the most important critical infrastructures: the smart grid. This chapter gives an overview of regulations and standards relevant to the smart grid as well as ongoing activities and standardization bodies.

7.1 Introduction

Today, the power market and the operation of power systems in general are strongly influenced by a large number of regulations and standards. Recently, many of these standards also have addressed information technology (IT) security as an important aspect of the protection of critical infrastructures. This chapter gives a (surely incomplete) overview of relevant regulation and standardization activities related to security of the smart grid. For a survey of proposed standardization activities related to the smart grid in general, the International Electrotechnical Commission (IEC) and National Institute of Standards and Technology (NIST) activities defining standardization road maps are referred to (the respective documents are referenced). Parts of this chapter have been taken from Fries and Hof[1] and the IEC.[2]

7.2 Standardization

The following sections provide a rough overview of the most important security-related standardization and regulation activities with respect to security for the smart grid. References to the original documents or further information are provided.

7.2.1 *International Organization for Standardization/ International Electrotechnical Commission*

The International Organization for Standardization (ISO) and the IEC are cooperating standardization bodies. The ISO provides international standards that target technical and organizational means in several application domains. The IEC develops international standards for all electrical, electronic, and related technologies. To deal with overlap between ISO and IEC, both standardization organizations cooperate in so-called joint technical committees.

7.2.2 *ISO/IEC 27000 Series*

The standard *Information Technology—Security Techniques—Information Security Management Systems"* consists of different parts. ISO/IEC 27001[3] specifies information security management requirements. The requirements are suited for use in certification. ISO/IEC 27002[4] provides the code of practice for information security management and establishes guidelines and general principles for initiating, implementing, maintaining, and improving information security management in an organization. ISO/IEC 27002 provides generic guidelines, which can be mapped to specific domains. This allows addressing specialties for the targeted application domain. One example is ISO 27011, targeting the mapping of ISO 27002 to the domain of telecommunication. A further example is provided by the German Deutsches Institut für Normung (DIN), which developed the DIN SPEC 27009[5] (cf. *Information Security Management Guidelines for Process Control Systems Used in the Energy Utility Industry on the Basis of ISO/IEC 27002*), mapping ISO 27002 guidelines and principles to the electric utility domain. This national specification has been submitted to the ISO for adopting the work to provide it as an international standard. This document is currently under evaluation, aiming at ISO 27019.

7.2.3 *IEC Smart Grid Strategic Group*

The IEC Smart Grid Strategic Group (SG3) has issued the *Smart Grid Standardization Roadmap* report (SMB/4175/R),[6] which encompasses requirements, status, and recommendations of standards relevant for the smart grid. A separate section of the *Smart Grid Standardization Roadmap* covers security-related topics. The report requests an overall security architecture coping with the complexity of smart grids. In addition, the following are recommendations pertaining to open items and necessary enhancements:

- A specification of a dedicated set of security controls (e.g., perimeter security, access control)
- A defined compartmentalization of smart grid applications (domains) based on clear network segmentation and functional zones
- A specification comprising identity establishment (based on trust levels) and identity management
- The necessity to consider security of the legacy components within standardization
- The harmonization with the IEC 62443 standard to achieve common industrial security standards
- Review, adapt, and enhance existing standards to support general and ubiquitous security across wired and wireless connections

7.2.4 ISO/IEC 62351-1 to 11

ISO/IEC 62351[7,8] is owned by the IEC Technical Committee 57 Working Group 15 (ISO/IEC TC 57 WG 15). Its scope is data and communication security for power system management and the associated information exchange between entities of the power system. ISO/IEC 62351 is used to establish and ensure end-to-end security. It is applied to protocols like IEC 61850, IEC 60870-x (energy automation), and ICCP (TASE.2, control center communication).

The standard has eight parts, each in a different state of completion. Further parts may be added in the future if necessary. The latest part targets the management of security credentials. Table 7.1 gives an overview of the parts of ISO/IEC 62351 and the current state of the standardization.

The first part of ISO/IEC 62351 introduces the standards and provides an overview. It addresses the security services needed in the

Table 7.1   Parts and Associated Standardization Status of the ISO/IEC 62351 Standard

IEC 62351	DEFINITION OF SECURITY SERVICES FOR	STANDARDIZATION STATUS
Part 1	Introduction and overview	Technical specification
Part 2	Glossary of terms	Technical specification
Part 3	Profiles including TCP/IP	Technical specification
Part 4	Profiles including MMS	Technical specification
Part 5	Security for IEC 60870-5 and derivatives	Technical specification
Part 6	Security for IEC 61850	Technical specification
Part 7	Network and system management (NSM) data object models	Technical specification
Part 8	Role-based access control for power systems management	Technical specification
Part 9	Credential management	Work in progress
Part 10	Security architecture guidelines	Technical report
Part 11	XML file security	New work item proposal

power domain. Part 2 provides the terminology used throughout the standard. Parts 3 to 8 are directly related to dedicated protocols typically used in energy automation, in particular ISO/IEC 61850 (IEC 62351-6) and ISO/IEC 60870-5-x (IEC 62351-5) as well as the mapping of those protocols to lower-layer protocols like the Transmission Control Protocol/Internet Protocol (TCP/IP) (IEC 62351-3) and Manufacturing Message Specification (MMS) (IEC 62351-4). The standard also addresses the mapping of security to the network management in part 7. For securing end-to-end communication, a broad range of cryptographic algorithms is used, including symmetric and asymmetric cryptographic algorithms to secure payloads and communication links. ISO/IEC 62351 does not try to reinvent the wheel. Hence, it uses well-known and widely used security protocols like TLS (Transport Layer Security). TLS offers security services like mutual authentication of communication peers as well as confidentiality and integrity protection of transmitted data. Among other attacks, this avoids man-in-the-middle attacks.

Part 3 of ISO/IEC 62351 defines security services for TCP/IP-based energy automation communication, including the specification of cipher suites (the allowed combination of encryption, authentication, and integrity protection algorithms) and requirements on certificates to be used for TLS. The definition of security services pays attention to characteristics of energy automation communication. For example, the definition of certificate revocation procedures is focused on the handling of CRLs (Certificate Revocation Lists), online validation of certificates (e.g., using the OCSP, Online Certificate Status Protocol) is not currently considered in edition 1 as communication links are severely limited in substations. Another characteristic of energy automation communication are long-lived connections. This requires the definition of strict key update and CRL update intervals to restrict the application of cryptographic keys not only for a dedicated number of packets but also for a dedicated time. Another challenge to consider is the interoperability requirements between the implementations of the products of different vendors. Nevertheless, TLS as underlying security protocol has evolved over time. Meanwhile its application is being recommended in substation automation. This drives the development of an edition 2 of part 3, which is currently under review. Edition 2 allows for using OCSP for certificate revocation as well as to better instrument TLS capabilities to cope with the target environment. Support of session resumption is just one example.

Part 4 of IEC 62351 specifies procedures, protocol enhancements, and algorithms targeting the increase of security messages transmitted over MMS. MMS is an international standard (ISO 9506) dealing with a messaging system for transferring real-time process data and supervisory control information either between networked devices or in communication with computer applications. Part 4 of IEC 62351 defines procedures on the transport layer, based on TLS, as well as on the application layer to protect the communicated information.

Part 5 of IEC 62351 defines additional security measures for serial communication. In particular, keyed hashes are used to protect the integrity of the data sent over a serial interface employing a symmetric key. This part also defines distinct key management for the use of keyed hashes. An edition 2 is expected soon, handling the update of update keys for the symmetric keys.

Part 6 of IEC 62351 defines security for IEC 61850 Peer-to-Peer Profiles. It covers the profiles in IEC 61850 that are not based on TCP/IP for the communication of GOOSEs (Generic Object Oriented Substation Events) and SVs (Sample Values) using, for example, plain Ethernet. This type of communication often uses multicast communication; each field device decides based on the message type and sender whether it processes the message. The security defined in part 6 uses digital signatures on the message level to protect the integrity of the messages. This approach is compatible with the use of multicast but requires a lot of computational power. Especially, the number of packets to be processed can be high. At a sample rate of 80 samples per power cycle, there are up to 4,000 packets per second for the common frequency of 50 Hz. Field devices used are typically not built to handle 4,000 signatures per second for generation or for verification. Hence, an edition 2 is targeted addressing this shortcoming. In the future, it is likely to use a group-based approach. Here, a group shares a symmetric key that is applied in the calculation of an integrity check value using keyed hash functions like AES-GMAC (Advanced Encryption Standard-Galois Message Authenticaton Code) or HMAC-SHA256 (Hash-based Message Authentication Code-Secure Hash Algorithm with key length 256). Digital signatures in this approach are only used to authenticate toward the key server distributing the group key.

Part 7 describes security-related data objects for end-to-end network and system management (NSM) and security problem detection. These data objects support the secure control of dedicated parts of the energy automation network. Part 7 can help to implement or extend intrusion detection systems for power system-specific objects and devices.

Part 8 supports role-based access control in terms of three profiles. Each of the profiles uses an own type of credential as there are identity certificates with role enhancements, attribute certificates, and software tokens. Role-based access control is necessary to support authorization in protection systems and in control center applications. Moreover, it supports stringent traceability. One usage example is the verification of who has authorized and performed a dedicated switching command.

Part 9 is a work in progress targeting the definition of key management supporting power system architectures in general and IEC 62351 specifically. It shall cover all key management-related parts of IEC 62351, helping to reuse key management options as much as possible, also in future parts to be defined.

Part 10 targets a technical report rather than a technical specification and provides an overview considering security for power system architectures. It motivates the incorporation of security right from the beginning and suggests certain security controls. The document is intended to foster the adaptation of security and thus does not provide a complete architecture but architecture elements. Moreover, it references several other documents providing comprehensive insight, like the NIST documents referenced previously.

Part 11 is currently a New Work Item Proposal targeting security for XML (eXtensible Markup Language) files. The goal of this part is the marking of information in messages and local data according to its sensitivity. This is necessary to allow a receiver of certain information to act on the information accordingly. This becomes especially evident if a receiver transforms and stores the information, which may later be queried by other applications.

A first glimpse at the current IEC 62351 parts shows that many of the technical security requirements to be applied to energy automation components and systems can be directly derived from the standard. For instance, parts 3 and 4 explicitly require the usage of TLS. They define cipher suites, which are to be supported as mandatory. These parts also define recommended cipher suites and deprecate cipher suites, which shall not be applied from the IEC 62351 point of view. Note that the mandatory cipher suites do not collapse with the cipher suites the different TLS versions (1.0, Request for Comments [RFC] 2246; 1.1, RFC 4346; 1.2, RFC 5246) stated as mandatory. IEC 62351 edition 1 standards always reference TLS version 1.0 to better address backward compatibility.

Analyzing the standard more deeply shows that several requirements are provided rather implicitly. These requirements are mostly related to the overall key management, which guarantees a smooth operation of the security mechanisms. IEC 62351 makes heavy use of certificates and associated private keys (e.g., when using TLS or GOOSE). However, key management is unspecified. Key management includes generation, provisioning, revocation, and initial distribution of keys and certificates to all related entities. It has been noticed that standardized key management is necessary for general operation as well as for the interoperability of the products of different vendors. This has been acknowledged and was the main reason to start working on part 9 as described.

Besides standard enhancements, which have become necessary through findings during the implementation of IEC 62351, new scenarios may also require the further evolvement of already-existing or new parts of the standard to better cope with new use cases.

7.2.5 ISO/IEC 62443

The ISO/IEC TC 65 WG 10 is currently standardizing ISO/IEC 62443,[9] targeting network and system security in industrial communication networks. ISO/IEC 62443 is a joint approach, together with International Society of Automation (ISA) 99 (see the next section); that is, ISA 99 documents will be submitted to the IEC voting process. The standard has different parts, which are in different states of completeness. *IEC 62443-1-1 (Terminology and Concepts), IEC 62443-2-1 (Establishment of an Industrial Automation and Control System [IACS] Security Program), and IEC 62443-3-1 (Security Technologies for IACS)* are currently available as standards. Work is ongoing on further parts addressing the definition of security levels, certification requirements, and the mapping of ISO 27002 to the industrial automation domain.

7.2.6 International Society of Automation

The ISA is a nonprofit society in the field of industry automation. Besides other duties, ISA is an important standardization body in the context of automation.

ISA-99 defines a framework addressing "Security for Industrial Automation and Control Systems."[10] This broad topic also includes energy automation. The framework covers processes for establishing an industrial automation and control system security program based on risk analysis, establishing awareness and countermeasures, and monitoring and cybersecurity management systems. It describes several categories of security technologies and the types of products available in those categories along with preliminary recommendations and guidance for using those security technologies. The standard consists of several subparts, which are in different states of completion.

7.2.7 Institute of Electrical and Electronics Engineers

The IEEE standard *IEEE 1686-2007: Standard for Substation Intelligent Electronic Devices (IEDs) Cyber Security Capabilities*[11] defines mandatory functions and features to accommodate critical infrastructure protection programs. It covers security in terms of access, operation, configuration, firewall revision, and data retrieval from IEDs. Encryption for the secure transmission of data, both within and external to the substation is not part of this standard.

Also applicable in the power system domain are the IEEE 802 standards:

- *IEEE 802.1X: Port Based Network Access Control* specifies port-based access control, allowing the restrictive access decisions to networks based on dedicated credentials. It defines the encapsulation of the EAP (Extensible Authentication Protocol) over IEEE 802, also known as EAP over local-area network (LAN) or EAPOL. The specification also includes key management, formally specified in IEEE 802.1AF.
- *IEEE 802.1AE: MAC* [Media Access Control] *Security* specifies security functionality in terms of connectionless data confidentiality and integrity for media access-independent protocols. It specifies a security frame format similar to Ethernet.
- *IEEE 802.1AR: Secure Device Identity* specifies unique per device identifiers and the management and cryptographic binding of a device to its identifiers.

7.2.8 International Council on Large Electronic Systems

The International Council on Large Electronic Systems (CIGRE) is an international organization covering technical, economic, environmental, organizational, ad regulatory aspects of electric power systems. The goals of CIGRE include providing state-of-the-art world practices to engineering personnel and specialists in the field.

7.2.9 Security for Information Systems and Intranets in the Electric Power System

CIGRE published the document, *Security for Information Systems and Intranets in Electric Power Systems.*[12] The guideline presents the work of the Joint Working Group D2/B3/C2-01, focusing on the importance of handling information security within an electric utility, dealing with various threats and vulnerabilities, the evolution of power utility information systems from isolated to fully integrated systems, the concept of using security domains for dealing with information security within an electric utility, and the use of the ISO/IEC 17799 standard (predecessor of ISO 27000).

7.2.10 *Treatment of Information Security for Electric Power Utilities*

Working Group D2.22 published the document, *Treatment of Information Security for Electric Power Utilities*. The document includes three reports:

- *Risk Assessment of Information and Communication Systems*[13]
- *Security Frameworks for Electric Power Utilities*[14] and
- *Security Technologies Guideline*[15]

The three reports provide practical guidelines and experiences for determining security risks in power systems and the development of frameworks, including control system security domains.

7.2.11 *North American Electric Reliability Corporation*

The mission of the North American Electric Reliability Corporation (NERC) is to ensure the reliability of the bulk power system in North America. To do so, NERC develops and enforces reliability standards and monitors users, owners, and operators for preparedness. NERC is a self-regulatory organization subject to oversight by the U.S. Federal Energy Regulatory Commission and governmental authorities in Canada. NERC has established the Critical Infrastructure Protection (CIP) Cyber Security Standards CIP-002 through CIP-011, which are defined to provide a foundation of sound security practices across the bulk power system. These standards are not designed to protect the system from specific and imminent threats. They apply to operators of bulk electric systems (BESs; see also North American Reliability Corporation[16]). The profiles originated in 2006. NERC CIP provides a consistent framework for security control perimeters and access management with incident reporting and recovery for critical cyber assets and cover functional as well as nonfunctional requirements. Table 7.2 provides an overview of the various NERC CIP parts.

The draft standard CIP-011 may not lead to new cybersecurity requirements but provides a new organization of the existing requirements of the existing CIP standards. New is the classification of BESs

Table 7.2   NERC CIP Parts

CIP	TITLE/CONTENT
002	*Critical Cyber Asset Identification*
	Identification and documentation of critical cyber assets using risk-based assessment methodologies
003	*Security Management Controls*
	Documentation and implementation of cybersecurity policy reflecting commitment and ability to secure critical cyber assets
004	*Personnel and Training*
	Maintenance and documentation of security awareness programs to ensure personnel knowledge on proven security practices
005	*Electronic Security Protection*
	Identification and protection of electronic security perimeters and their access points surrounding critical cyber assets
006	*Physical Security Program*
	Creation and maintenance of physical security controls, including processes, tools, and procedures to monitor perimeter access
007	*Systems Security Management*
	Definition and maintenance of methods, procedures, and processes to secure cyber assets within the electronic security perimeter to not adversely affect existing cybersecurity controls
008	*Incident Reporting and Response Planning*
	Development and maintenance of a cybersecurity incident response plan that addresses classification, response actions, and reporting
009	*Recovery Plans for Critical Cyber Assets*
	Creation and review of recovery plans for critical cyber assets
010	*Bulk Electrical System Cyber System Categorization (draft)*
	Categorization of BES systems that execute or enable functions essential to reliable operation of the BES into three different classes
011	*Bulk Electrical System Cyber System Protection (draft)*
	Mapping of security requirements to BES system categories defined in CIP-010

into the three categories—low-, medium-, and high-impact BES cybersystems—and their mapping to security controls. Currently, work is ongoing on version 5 of the set of NERC CIP documents.

7.2.12 Internet Engineering Task Force

The Internet Engineering Task Force (IETF) develops international standards targeting protocol suites operating on different layers of the Open System Interconnection (OSI) stack. Prominent examples of standards relate to TCP/IP and the IP suite. The IETF cooperates also with other standardization bodies, like the ISO/IEC or W3C (World Wide Web Consortium). The following RFCs are applicable in the power system domain and therefore stated here:

- The IETF published RFC 6272, *Internet Protocols for the Smart Grid*,[17] which contains an overview of security considerations and a fairly thorough list of potentially applicable security technology defined by the IETF.
- *RFC 3711: Secure Real-Time Transport Protocol (SRTP)*[18] may be used for securing Voice over Internet Protocol (VoIP) communication, including video conferencing or video surveillance.
- RFC 4101,[19] RFC 4102,[20] RFC 4103[21] are the base standards for IP Security (IPSec) providing layer 3 security, typically used for virtual private networks (VPNs) or for remote access. The listed RFCs describe general architecture as well as the two modes AH (Authentication Header) and ESP (Encapsulated Security Payload).
- *RFC 4962: Authentication, Authorization, and Accounting*[22] provides guidance for authentication, authorization, and accounting (AAA) key management and an architecture allowing centralized control of AAA functionality.
- *RFC 5246: Transport Layer Security (TLS)*[23] provides layer 4 security for TCP/IP-based communication, currently used in IEC 62351. Note that there are several extensions to TLS for additional cipher suites, transmission of additional information like authorizations or OCSP responses, and so on. These extensions are not listed here explicitly.
- *RFC 5247: Extensible Authentication Protocol (EAP)*[24] provides a key management framework for EAP. Single EAP methods are defined in separate RFCs. EAP is typically used for controlling device (or human) access to networks.
- *RFC 5746: Datagram Transport Layer Security (DTLS)*[25] provides layer 4 security for communication based on the User Datagram Protocol (UDP)/IP. It may be applied in scenarios for which TLS is not applicable.
- *RFC 6407: Group Domain of Interpretation (GDOI)*[26] defines group-based key management, currently used in IEC 61850-90-5.

This list states the most obvious standards to be used but is not limited to them.

7.3 National Regulations

Besides international standardization bodies and activities, many national organizations and activities influence the development of energy automation systems in the respective countries. This section covers national activities in the United States and Germany as well as activities on a European level.

7.3.1 National Institute of Standards and Technology

The NIST is a U.S. federal technology agency that develops and promotes measurement, standards, and technology. The following NIST documents cover security in energy automation systems or can be directly applied to security in the smart grid.

7.3.2 Special Publication 800-53

NIST Special Publication (SP) 800-53, *Recommended Security Controls for Federal Information Systems*[27] provides guidelines for selecting and specifying technical and organizational security controls and connected processes for information systems supporting the executive agencies of the federal government to meet the requirements of Federal Information Processing Standard (FIPS) 200 (*Minimum Security Requirements for Federal Information and Information Systems*).[28] It provides an extensive catalog of security controls and maps these in a dedicated appendix to industrial control systems (ICSs).

7.3.3 Special Publication 800-82

NIST SP 800-82: *Guide to Industrial Control Systems (ICS) Security*[29] covers how to secure ICSs, including supervisory control and data acquisition (SCADA) systems, distributed control systems (DCSs), and other control system configurations, such as programmable logic controllers (PLCs). It uses the NIST SP 800-53 as a basis and provides specific guidance on the application of the security controls in

NIST SP 800-53. This publication is an update to the second public draft, which was released in 2007.

7.3.4 Special Publication 1108

NIST SP 1108, *NIST Framework and Roadmap for Smart Grid Interoperability Standards*[30] describes a high-level conceptual reference model for the smart grid. It lists 75 existing standards applicable or likely to be applicable to the ongoing development of the smart grid. The document also identifies future issues, including 15 high-priority gaps and potential harmonization issues for which new or revised standards and requirements are needed.

7.3.5 NIST IR 7628

NIST IR 7628[31,32] originates from the Smart Grid Interoperability Panel (Cyber Security WG) and targets the development of a comprehensive set of cybersecurity requirements building on the NIST SP 1108, also stated previously. The document consists of three subdocuments targeting strategy,[30] security architecture[31] and requirements, and supportive analyses and references.[33]

7.3.6 U.S. Department of Homeland Security

The *Catalog of Control Systems Security—Recommendations for Standards Developers*[34] of the U.S. Department of Homeland Security summarizes practices of various industry bodies to increase the security level of control systems both from physical and from cyber attacks. The catalog is not limited to energy automation but may also be used for other domains to develop a cybersecurity program.

7.3.7 Bundesverband für Energie- und Wasserwirtschaft—BDEW (Germany)

The German Bundesverband für Energie- und Wasserwirtschaft—BDEW was founded by the federation of four German energy-related associations: Bundesverband der deutschen Gas- und Wasserwirtschaft (BGW), Verband der Verbundunternehmen und Regionalen Energieversorger in Deutschland (VRE), Verband der Netzbetreiber

(VDN), and Verband der Elektrizitätswirtschaft (VDEW). The BDEW published a white paper[35] defining basic security measures and requirements for IT-based control, automation, and telecommunication systems, taking into account general technical and operational conditions. It can be seen as a further national approach targeting similar goals as NERC CIP. The white paper addressed requirements for vendors and manufacturers of power system management systems and is used as an amendment to tender specification.

7.3.8 European Union's Task Force Smart Grid

Within the European Union, a dedicated expert group of the Task Force Smart Grid is currently working on regulatory recommendations for data safety data handling and data protection.[36] The goal of the task force is the identification and production of a set of regulatory recommendations to ensure EU-wide consistent and fast implementation of smart grids while achieving the expected smart grids' services and benefits for all users involved. The goal of the expert group for security is the identification of an appropriate regulatory scenario and recommendations for data handling, security, and consumer protection to establish a data privacy and data security framework that both protect and enable.

7.3.9 Results from the European Smart Grid Coordination Group

The objective of mandate M/490[37] is the development or update of a set of consistent standards within a common European framework that will facilitate the implementation of the different high-level smart grid services and functionalities. The Smart Grid Coordination Group (SGCG) was founded in June 2011 to directly address mandate M/490. It is a joint activity from CEN (European Committee for Standardization, http://www.cen.eu), CENELEC (European Committee for Electrotechnical Standardization, http://www.cenelec.eu/), and ETSI (European Telecommunications Standards Institute, www.etsi.org/) to run for almost 2 years until the end of 2012 resulting in 4 reports. The activity will be enhanced for another 2-year period.[38] As security is one of the targets of this mandate, a dedicated

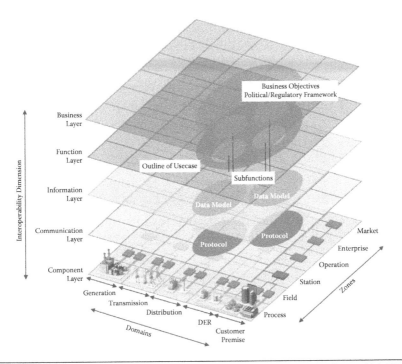

Figure 7.1 Smart Grid Architecture Model[39] (SGAM, cf. Smart Grid Coordination Group, *Joint CEN, CENELEC and ETSI Activity on Standards for Smart Grids*, 2009[38]). DER = Distributed Energy Resources.

subgroup—the Smart Grid Information Security (SGIS)—addresses this topic explicitly.

IT security is closely connected to the architectural model provided by the reference architecture group as the SGAM (Smart Grid Architectural Model). This model is presented as a cube in which use cases can be mapped on zones and domains on different layers as depicted in Figure 7.1.[39]

Security applies basically to every interface and component in the SGAM depending on the intended use cases. To provide guidance regarding which security means are to be applied, an analysis of the specific use cases is necessary. This is typically being done by performing a threat and risk analysis for a dedicated scenario targeting the identification of potential vulnerabilities based on the analysis of the considered scenario or use case. Based on this analysis, security requirements can be derived and appropriate countermeasures can be recommended. The security working group has developed a methodology for this approach, which is described in the WG report as the security tool box.

This is also supported by some other work, which has been done by providing a mapping of security requirements provided by the NIST IR 7628 (see previous discussion) to standards like ISO 27001 or IEC 62351. This mapping has also been done regarding the NERC CIP documents. The goal of this mapping was the identification of gaps, which need to be addressed by the responsible standardization body. There have been explicit comments on IEC 62351 on the technical side and a clear push for the technology covered. Moreover, the German DIN SPEC 27009 (see previous discussion) has been pushed toward internationalization in ISO for the organizational security part in the energy utility industry. Meanwhile this activity resulted in ISO TR 27019.

It has been acknowledged that the work of the Smart Grid Coordination Group will not end with the working period on the M/490 mandate. It is expected that there will be a refocusing of the group to address specific issues discovered during the first 2-year runtime.

7.4 Summary

The maternity of selected standards and their applicability is presented in Figure 7.2 as proposed in the European-funded project ESCORTS (Efficient Solar Cells based on Organic and Hybrid Technology).[40] The figure is intended to provide a better overview for operator and manufacturer regarding which standard influences their

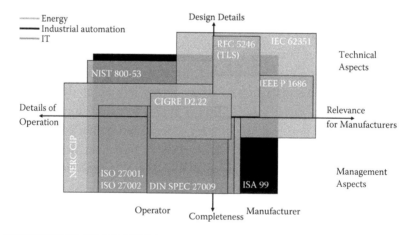

Figure 7.2   Scope and completeness of selected standards (enhanced version of ESCORTS Project[40]).

business most. The different shades of grey in the figure indicate the targeted audience.

While IEC 62351 addresses the energy sector, more specifically substation automation systems, NERC CIP generally targets energy operators. While ISO 27000 and NIST 800-53 are mainly targeted to IT environments (thus targeted at protecting information), other standards, such as ISA 99 or IEEE P1686, directly address (industrial) automation systems. It should be noted that NIST SP 800-53 Appendix I is explicitly for ICSs, as is NIST SP 800-82.

Standards extending to the right in the x-axis direction in the figure have relevance for manufacturers. Typically, such standards have detailed technical requirements up to the definition of special security protocols, which must be implemented by the manufacturers. In contrast, the more a standard extends to the left of the x axis, the more it is focused on a secure operation. NERC CIP, for example, prescribes specific actions for operators to do, thus providing implicit requirements to the manufacturers to support the operators.

Standards extending to the top of the y axis list precise design details and leave little room for interpretation. IEC 62351, for instance, provides design details to such an extent that device interoperability between various manufacturers can be guaranteed.

Standards extending to the bottom of the y axis cover a broad range of various security areas and thus can be consulted to obtain an estimation of the overall security level.

A smart grid information infrastructure can be characterized as a complex, heterogeneous interconnected system involving different usages, stakeholders, and technologies. This chapter gave an overview of smart grid standardization, regulation, and guideline activities.

Besides the stated activities in Germany and North America, there are further activities, like the road map activities in Asia (especially in Japan and China), addressing smart grid use cases and connected standardization.

Several properties of a smart grid pose challenges for designing a practically deployable and usable security solution for the smart grid. One point is the long lifetime of energy devices compared to the lifetime of IT equipment. Devices once deployed will remain in the field for many years until replacement. A security design has to consider migration aspects to cope with legacy devices, and it has to be designed with the expectation to be adequate for many years. The huge number of heterogeneous devices requires a practical, low-effort or zero-effort management of cryptographic keys and certificates. The diversity of devices, use cases, and stakeholders implies that different kinds of security domains have to be supported within a smart grid. Further challenges are posed through the necessary coordination and alignment of requirements from a plurality of stakeholders (operator, product vendors, consumers, regulations, etc.).

One base for broad adaptation of security as a system inherent feature is also the interoperability between different vendor's products. This is provided by standardization.

References

1. Steffen Fries and Hans-Joachim Hof, Security considerations in the smart grid, in: Lars Berger and Kris Iniewski, *Smart Grids*, Wiley, New York, May 2012.
2. International Electrotechnical Commission, *IEC 62351-10 TR, Security Architecture Guidelines for TC57 Systems*, IEC, Geneva, Switzerland, October 2012.
3. International Organization for Standardization, ISO 27001, ISO/IEC 27001:2005 Information technology—Security techniques—Information Security Management Systems—Requirements, http://www.iso27001security.com/html/27001.html
4. International Organization for Standardization, *ISO 27002, ISO/IEC 27002: 2005 Information Technology—Security Techniques—Information Security Management Systems—Code of Practice for Information Security Management*, http://www.iso27001security.com/html/27002.html
5. *DIN Spec 27009, Information Security Management Guidelines for Process Control Systems Used in the Energy Utility Industry on the Basis of ISO/IEC 27002*, March 2012.
6. IEC Smart Grid Strategic Group (SG3), Smart Grid Standardization Roadmap http://www.iec.ch/cgi-bin/restricted/getfile.pl/SMB_4175e_R.pdf?dir=SMB&format=pdf&type=_R&file=4175e.pdf
7. ISO-IEC 62351, Part 1-11, http://www.iec.ch/cgi-bin/procgi.pl/www/iecwww.p?wwwlang=E&wwwprog=sea22.p&search=iecnumber&header=IEC&pubno=62351&part=&se=
8. Steffen Fries, Hans Joachim Hof, Thierry Dufaure, and Maik Seewald, Security for the smart grid—enhancing IEC 62351 to improve security in energy automation control, International Journal on Advances in Security. April 2011, http://www.thinkmind.org/download.php?articleid=sec_v3_n34_2010_7

9. ISO-IEC 62443, Part 1-3, http://www.iec.ch/cgi-bin/procgi.pl/www/iecwww.p?wwwlang=E&wwwprog=sea22.p&search=iecnumber&header=IEC&pubno=62443&part=&se=

10. International Society of Automation, *ISA 99 Industrial Automation and Control Systems Security, Standards Framework,* http://www.isa-99.com/.

11. *ISO-IEC IEC 62357, Part 1: Reference Architecture for TC57,* Second Draft, July 2009.

12. CIGRE Joint Working Group D2/B3/C2-01, Managing Information Security in an Electric Utility, http://d2.cigre.org/content/download/11370/334067/version/2/file/Managing+information+security+in+an+Electric+utilityID41VER28.pdf

13. CIGRE (International Council on Large Electronic Systems) Working Group D2.22 report, *Risk Assessment of Information and Communication Systems,* August 2008, Electra.

14. CIGRE report, *Security Frameworks for Electric Power Utilities,* WG D2.22, December 2008, Electra.

15. CIGRE report, *Security Technologies Guideline,* WG D2.22, June 2009, Electra.

16. North American Reliability Corporation, Standards: Reliability Standards, http://www.nerc.com/page.php?cid=2|20.

17. http://tools.ietf.org/html/rfc6272

18. http://tools.ietf.org/html/rfc3711

19. http://tools.ietf.org/html/rfc4101

20. http://tools.ietf.org/html/rfc4102

21. http://tools.ietf.org/html/rfc4103

22. http://tools.ietf.org/html/rfc4962

23. http://tools.ietf.org/html/rfc5246

24. http://tools.ietf.org/html/rfc5247

25. http://tools.ietf.org/html/rfc5746

26. http://tools.ietf.org/html/rfc6407

27. National Institute of Standards and Technology, *NIST SP 800-53, Recommended Security Controls for Federal Information Systems and Organizations,* Revision 3, August 2009, http://csrc.nist.gov/publications/nistpubs/800-53-Rev3/sp800-53-rev3-final.pdf

28. Federal Information Processing Standard (FIPS) 200: Minimum Security Requirements for Federal Information and Information Systems http://csrc.nist.gov/publications/fips/fips200/FIPS-200-final-march.pdf

29. National Institute of Standards and Technology, *NIST SP 800-82, Guide to Industrial Control Systems (ICS) Security,* Draft, September 2008, http://csrc.nist.gov/publications/drafts/800-82/draft_sp800-82-fpd.pdf

30. National Institute of Standards and Technology, *NIST Framework and Roadmap for Smart Grid Interoperability Standards,* Version 1.0, January 2010, http://www.nist.gov/public_affairs/releases/upload/smartgrid_interoperability_final.pdf

31. National Institute of Standards and Technology, *NIST IR 7628 Guidelines for Smart Grid Cyber Security, Vol. 1 Smart Grid Cyber Security Strategy*, Draft, July 2010, http://csrc.nist.gov/publications/PubsDrafts.html#NIST-IR-7628

32. National Institute of Standards and Technology, *NIST IR 7628 Guidelines for Smart Grid Cyber Security, Vol. 3 Supportive Analyses and References*, Draft, July 2010, http://csrc.nist.gov/publications/PubsDrafts.html#NIST-IR-7628

33. National Institute of Standards and Technology, *NIST IR 7628 Guidelines for Smart Grid Cyber Security, Vol. 2 Security Architecture and Security Requirements*, Draft, July 2010, http://csrc.nist.gov/publications/PubsDrafts.html#NIST-IR-7628

34. U.S. Department of Homeland Security, *Catalog of Control Systems Security—Recommendations for Standards Developers*, June 2010, http://www.us-cert.gov/control_systems/pdf/Catalog%20of%20Control%20Systems%20Security%20-%20Recommendations%20for%20Standards%20Developers%20June-2010.pdf

35. BDEW—Bundesverband der Energie- und Wasserwirtschaft, *Datensicherheit*, http://www.bdew.de/bdew.nsf/id/DE_Datensicherheit

36. EU Task Force Smart Grid, Expert Group 2, *Regulatory Recommendations for Data Safety Data Handling and Data Protection*, February 16, 2011, http://ec.europa.eu/energy/gas_electricity/smartgrids/doc/expert_group2.pdf

37. European Commission, Directorate-General for Energy, *M/490, Standardization Mandate to European Standardisation Organisations (ESOs) to Support European Smart Grid Deployment*, March 2011, http://ec.europa.eu/energy/gas_electricity/smartgrids/doc/2011_03_01_mandate_m490_en.pdf.

38. Smart Grid Coordination Group, *Joint CEN, CENELEC and ETSI Activity on Standards for Smart Grids*, 2009, http://www.cen.eu/cen/Sectors/Sectors/UtilitiesAndEnergy/SmartGrids/Pages/default.aspx

39. Siemens, Siemens Develops European Architecture Model for Smart Grid, http://www.siemens.com/press/en/pressrelease/?press=/en/pressrelease/2012/infrastructure-cities/smart-grid/icsg201205018.htm.

40. ESCORTS Project, Home page, http://www.escort-project.eu/.

8

Vulnerability Assessment for Substation Automation Systems

ADAM HAHN, MANIMARAN GOVINDARASU, AND CHEN-CHING LIU

Contents

Growing cybersecurity concerns within the smart grid have created increasing demands for vulnerability assessments to ensure adequate cyber protections. This chapter reviews vulnerability assessment requirements within substation automation communication and computation mechanisms and identifies a

227

methodology to evaluate security concerns while avoiding any negative impact on operational systems. Finally, national and industry efforts to expand assessment capabilities within this domain are addressed.

8.1 Introduction

The smart grid creates an increasing dependency on the cyber infrastructure to monitor and control the physical system. While supervisory control and data acquisition (SCADA) technology has been utilized for many years, the increasing interconnectivity expands the general cyberattack surface. Recent government reports have raised concerns about the general security posture of these systems.[1,2] In an attempt to mitigate these concerns, the North American Reliability Corporation (NERC) has produced compliance requirements for critical cyber resources to ensure an appropriate protection level.[3] These documents specifically require that a cyber vulnerability assessment is performed to verify that they meet the appropriate security requirements. Unfortunately, the vulnerability assessment process is not well understood for this domain due to numerous constraining properties, including

- Heavy reliance on undocumented, proprietary communication protocols.
- High availability requirements that limit testing of operational systems.
- Software platforms that have not undergone a thorough security analysis and have not been engineered to undergo a security review.
- Geographic distribution of resources limiting physical resource accessibility.

Figure 8.1 provides an overview of the communication infrastructure within the smart grid. Distribution, transmission, and generation domains are identified as well as their interconnectivity and dependency on other parties. The figure identifies various protocols necessary to support this communication and highlights the connectivity between substations and control centers. Security concerns

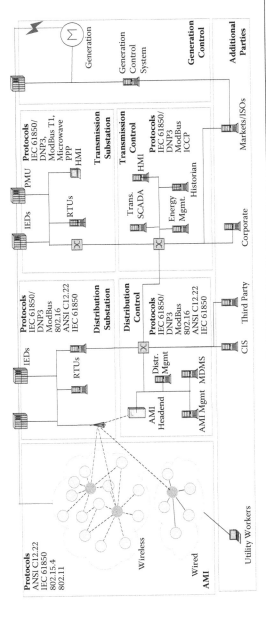

Figure 8.1 Smart grid environment. ANSI = American National Standards Institute; CIC= Customer Information System; HMI = Human Machine Interface; ICCP = Inter-Control Center Protocol; IED = Intelligent Electronic Device; ISO = Independent System Operator; MDMS = Meter Data Management System; PMU = Phasor Measurement Unit; PPP = Point-to-Point Protocol; RF = Radio Frequency; RTU = Remote Terminal Unit

are specifically presented by the unprotected substations and feasible external accessibility of control centers due to corporate and vendor requirements. In addition, smart grid advancements such as advanced metering infrastructures (AMIs) and wide-area measurement systems (WAMSs) will only present greater interconnectivity of these systems.

This chapter addresses concerns for performing a comprehensive vulnerability assessment within this domain based on the previous constraints. A methodology is presented to appropriately structure assessment efforts. Software tools to assist in the evaluation process are introduced, and their application to this domain is reviewed. In addition, current efforts to expand assessment capabilities are introduced.

8.2 Assessment Methodologies

A strong methodology is imperative to ensure that testing efforts appropriately target the technologies involved within the environment and likely threats to the system. Security testing efforts can be tailored toward different objectives based on the intended scope. The development of vulnerability assessment methodologies has been well explored within traditional information technology (IT) environments; the following list provides some examples:

- *National Institute of Standards and Technology Special Publication (NIST SP) 800-115, Technical Guide to Information Security Testing and Assessment*[4]
- *NIST SP 800-53A, Guide for Assessing the Security Controls in Federal Information Systems*[5]
- *Open Source Security Testing Methodology Manual (OSSTMM)*[6]

A high-availability environment such as the smart grid presents a requirement for nonintrusive methodologies. Activities that could potentially cause availability or integrity problems must be restricted. This chapter presents an example of methodology based on that proposed in NIST 800-115, but with specific tailoring to avoid availability concerns. Figure 8.2 provides an overview of the major steps, specifically planning, execution, and postexecution. This chapter primarily highlights the execution phase as it typically involves most of the technical issues. The main components of the execution phase are (1) review techniques, (2) target identification and analysis,

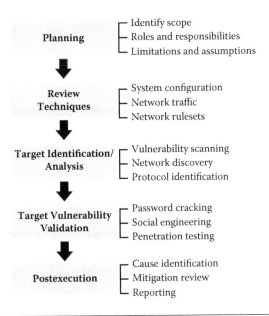

Figure 8.2   Vulnerability assessment plan.

and (3) target vulnerability validation. These are further explained in the following sections.

8.2.1 Planning

A key component of the planning phase is the scoping and monitoring of testing activities to ensure they do not negatively interfere with normal operations. This should involve establishing a representative test environment that maintains similar configurations. While assessment scope could vary based on the assessment's intent, the NERC CIP (North American Reliability Corporation Critical Infrastructure Protection) focused assessments on the control centers, substations, and associated communications.[7] Specific concerns within these components are identified next.

8.2.1.1 Control Center

Control centers will typically contain sets of operator/engineering workstations, control servers, and the resulting network infrastructure. This environment will likely resemble a traditional IT system containing Windows/Unix systems and similar networking switches/routers. While the control system software will be specific to the power domain, other supporting services such as Web servers, authentication services (Lightweight Directory Access Protocol (LDAP), Active Directory) and databases may be used. Specific systems within this environment include

- SCADA/EMS (energy management system) servers: Control servers that perform monitoring, control, and state estimation tasks
- Historians: Databases that maintain historic control system data for trending analysis
- Human-machine interfaces (HMIs): Systems that provide operator interfaces to the SCADA/EMS systems

Often, control systems maintain some connectivity to other corporate local-area networks (LANs) or other third parties due to requirements to collect operational data or provide vendor access.[8] The high-security requirements of this environment strongly emphasize scrutiny over remote access capabilities. In addition, while authentication and authorization present key security mechanisms, it must be assumed that in emergency situations, these controls may require some override function.

Assessment Guidance: Specific security concerns with the control environment include (1) appropriate network segregations through routing and firewall rules; (2) implementations of demilitarized zone (DMZ) for services needing access by both control and corporate environments; (3) appropriate patching and system configurations; and (4) sufficient authentication and authorization enforcement.

8.2.1.2 Substations   Substations within both the transmission and distribution domain have unique security requirements due to their geographic location. The communication links provide a specific concern due to the criticality of the transmitted data and their heavy use of wireless communication. All communication paths between the control center and substation, along with all intersubstation communications, require thorough analysis. *Field devices* are the components that perform the actual sensing and actuation functions throughout the grid. The term *field devices* is usually a generalization of various devices, including intelligent electronic devices (IEDs), programmable logic controllers (PLCs), and remote terminal units (RTUs). Typically, these are embedded systems with limited processing capabilities, nonstandard operating systems, and software platforms. This increases the likelihood of vulnerabilities and creates difficulties during the assessment process. Often, these devices are not Internet Protocol (IP) enabled, and if they are, they may implement incomplete or frail networking stacks that limit analysis capabilities.

Assessment Guidance: Specific security concerns with substation environments include (1) identification of all field device networking capabilities; (2) sufficient authentication of all accessible field device management/administrative functions; (3) cryptographically protected network communication between control centers or other substations; and (4) auditing of control/monitoring functions, authentication attempts, and device reconfigurations.

8.2.1.3 Network Protocol Overview   Protocols used within a control system vary from those commonly found in traditional IT environments. They are primarily responsible for transmitting binary and analog values on periodic intervals between systems. In addition, many of these protocols were designed and deployed before the proliferation of modern cybersecurity concerns. This section introduces numerous communication protocols, provides a brief explanation, and then identifies necessary security concerns that require inspection during the assessment.

8.2.1.3.1 Distributed Network Protocol   The Distributed Network Protocol (DNP3) is commonly used within the electric grid, especially in substation automation. While DNP3 has been used for many years, it was recently adopted as an Institute of Electrical and Electronics Engineers (IEEE) standard (IEEE Standard 1815).[9] The protocol operates in a master/slave paradigm; the master is typically represented by the control server or RTU, and the slave functions as the field device or *outstation*. With this model, the master is able to transmit commands and receive readings from the various field units.

While packets are encapsulated with their own data, transport, and application layers, the application layer plays the most important role in the assessment process. Each command and response is encapsulated within a DNP application service data unit (ASDU). The ASDU contains a *function code* used to identify the purpose of the message (e.g., read, write, confirm, response). The function code is then followed by one or more *objects* that identify the *data type* and value associated with the function code. Data types are typically analogs of digital inputs/outputs.

Authentication within DNP3 is enforced by categorizing function codes as *critical* and *noncritical*. Critical functions are typically those that perform some control or initiate a change on the outstation. Critical functions differ from noncritical in that the outstation can require a hash message authentication code (HMAC). A HMAC uses a shared key combined and a message hash to verify the message's authenticity and integrity. The HMAC calculation is based on the following set of preshared keys:

- Control key, to authenticate messages sent by the master
- Monitoring key, to authenticate messages sent by the outstation
- Update key, to perform a secure key update for both the control and monitoring keys

In addition to the traditional utilization of DNP3, additional work reviewed the use of Transport Layer Security (TLS) or Internet Protocol Security (IPSec) to provide a stronger underlying layer of security.[10]

Assessment Guidance: A secure implementation of the DNP3 protocols should achieve the following objectives: (1) identification of the communication path for all DNP3 traffic; (2) identification of all functions/objects that require authentication; (3) verification of the appropriate authentication on the resulting commands/responses; (4) identification of all communications protected by other means (e.g., IPSec virtual private networks [VPNs]); and (5) analysis of the key update exchanges.

8.2.1.3.2 International Electrotechnical Commission 61850   The transition to a smarter electric grid has required the development of more dynamics protocols. International Electrotechnical Commission (IEC) 61850 has been developed to provide increased interoperability, specifically in substation automation, and provides improved support of security mechanisms such as authentication and encryption. IEC

61850 presents an object-oriented approach to identifying substation components to simplify configuration and interoperability. Each *physical device* within the substation is represented by an IEC 61860 *object*; this object can then have sublogical devices, *logical node, data,* and *data attributes*. Nodes are assigned names based on their function; for example, logical node MMXU is used for a measurement, while XCBR is used for a circuit breaker. This naming scheme makes network traffic analysis more intuitive.

IEC 61850 is a complex protocol capable of sending various message types, including Generic Object Oriented Substation Event (GOOSE), Generic Substation Status Event (GSSE), and Sample Measured Values (SMVs). This chapter focuses on GOOSE as its utilization is more prevalent.

GOOSE relies on Ethernet virtual local-area networks (VLANs) (802.1q) to perform multicast delivery of content within a 4-ms time frame as required for protective relaying within substations. GOOSE messages can enable digital signatures to both authenticate and ensure the integrity of received messages. However, since digital signatures are based on public key cryptography and certificates, some certificate management function must be deployed. This distribution of certificates and the utilization of certificate authorities (CAs) become critical to understanding the security of the resulting IEC 68150 communications.

Assessment Guidance: A secure implementation of IEC 61850 should achieve the following objectives: (1) identification of the communication path for all traffic; (2) identification of the use of digital signatures or encryption; (3) identification of the VLAN 802.1q configuration on the network device for accurate inclusion of necessary systems and appropriate device configuration; and (4) a review of certificate distribution and trusts of CAs.

8.2.1.4 Supporting Protocols   Many common IT protocols are found within control systems and introduce security concerns. Domain Name System (DNS) is frequently used but can be problematic due to its dependency on Internet access as it may provide a covert channel for attackers.[11] DNS's utilization should be reviewed to ensure it does not introduce unnecessary external access points.

The Simple Network Management Protocol (SNMP) is often used by various devices within control systems to perform device administration. Access to SNMP configuration is protected by secret *community strings*; however, default strings such as "public" and "private" are often not changed. The use of a default community string should be reviewed, specifically those that allow write access to devices.

8.2.2 Review Techniques

The review step specifically addresses any nonintrusive analysis of data that can be obtained from systems and networks. These activities include system configuration documents/files, network device configuration/rule sets, and network traffic. Review techniques will play a critical role in the assessment process for the power grid as they are significantly less likely to have an impact on system operations.

8.2.2.1 System Configuration Review   Reviewing system configurations provides a nonintrusive method of determining potential vulnerabilities. Traditionally, this involves the review of any configuration files and the execution of commands that provide current system status. This information can then be correlated with any known secure baselines for the system to determine potential vulnerabilities. This review type is most effective when system configurations are well known. While this is typically the case with popular operating systems and network services, information is often unavailable for the software platforms and field devices used to support the grid. Research into the identification of secure software platform configurations has been explored by the Bandolier project.[12] This effort reviews popular software with the electric grid and establishes assessment capabilities based on other popular assessment tools (e.g., OVAL and Nessus).

8.2.2.2 Network Configurations/Rule Sets   Determining the network architecture is an important aspect of the security assessment process. This step focuses on the review of network device configurations to ensure they appropriately enforce the desired network architecture. This step is critical within the SCADA paradigm due to a heavy reliance on a secure network perimeter.[3] Incorrect assumptions about networking configuration may provide access to unauthorized users, which is specifically concerning due to weak authorization capabilities within many of the field devices.

Tools to assist in the review of network configurations and firewall rule sets are critical to the assessment process due to their relative difficulty of interpretation and the heavy interconnectivity between various devices. Fortunately, some tools have been developed to assist in this task. The Network Access Policy Tool (NetAPT) is the result of research efforts to automate the interpretation of network configurations and verify that they meet some previously assumed network policy.[13]

Future research should expand current tools to incorporate increased understanding of control system communication protocols and network topologies to provide an increased context for configuration analysis.

8.2.2.3 Network Traffic Review   Network traffic review provides a method to do *passive discovery* of the various network communications. This provides the assessor with an understanding of many systems, ports, and protocols being used within the environment. It also provides the ability to analyze various security-related information, such as whether encryption and authentication are being used appropriately.

There are various software tools available to perform network sniffing. Wireshark is an open source packet sniffer that maintains protocol dissectors for most popular IT and SCADA protocols, including DNP, IEC 61850, ModBus, and object linking and embedding (OLE) for process control (OPC).[14] While Wireshark provides strong functionality, more advanced tools have been developed to assist in this process. One particular tool, Sophia, is being developed by Idaho National Lab to utilize network discovery capabilities to identify the network communications.[15] Sophia uses network monitoring to determine the current architecture and communication requirements and identify any anomalies within the environment.

While network traffic review is necessary to understand the system and services operating on the network, it does not provide sufficient analysis of the network activity. Various systems or services may perform only transient communications and may not be detected through the sniffing. In addition, not all service configurations can be accurately

Table 8.1 System Configuration Review Tools

TOOL	TARGETED VULNERABILITIES	NEGATIVE IMPACT	DOMAIN SUPPORT
Bandolier	SCADA software configurations	Low	Full
NetAPT	Firewall rule set configurations	None	Full
Wireshark	Networking configuration and authentication/encryption verification	Low	Full
Sophia	Networking configuration and authentication/encryption verification	Low	Full

extracted from the communications, especially if the traffic is encrypted or the protocol's format is not well known. In these cases, additional activities must be performed to provide an accurate system view.

Table 8.1 presents an overview of the presented tools necessary to support the review techniques documented in this section. The table documents vulnerabilities that the tool can help discover, its ability to negatively impact operational systems, and how well it supports smart grid environments.

8.2.3 Target Identification and Analysis

After the initial review steps, a more in-depth analysis of specific components should be performed for target identification and analysis. Often, these activities can be considered intrusive since they require transmitting various requests to systems in an attempt to identify system configurations. These activities could have a negative impact on operational systems and ideally should be performed on a representative test environment.

8.2.3.1 Network Discovery Network discovery traditionally involves probing the various addresses on the system to discover all operating systems and services. The discovery phase typically uses various types of scanning tools that can send various probe packets in the network and interpret the responses to identify operating services. This activity, referred to as *port scanning*, uses ICMP (Internet Message Control Protocol) scans to determine active systems while using Transmission Control Protocol/User Datagram Protocol (TCP/UDP) scans to identify open ports.

A popular port-scanning tool, NMap, provides many different network probe types and reporting capabilities.[16] The tool's scanning capabilities include ICMP, ARP (Address Resolution Protocol), UDP, and numerous TCP scans with various flag configurations. NMap maintains a dictionary of known port/protocol mapping to help identify operating services as well as an operating system detection feature that may be useful when analyzing field devices for which little system information is known.

8.2.3.2 Vulnerability Scanning Vulnerability-scanning techniques have traditionally utilized network inspection methods to evaluate operating systems and network services in an attempt to identify vulnerabilities. This technique depends on a database of known vulnerability fingerprints that can be identified by various network probes. Vulnerability scanning can be an effective way to determine unpatched software and default/insecure configurations. While vulnerability-scanning tools remain popular due to their ability to inspect full ranges of systems and services, they may not be appropriate for an operational environment due to previously addressed availability and integrity concern. In addition, since this technique is limited to network probing, the amount of collectible information is limited.

Nessus is a popular vulnerability-scanning tool that is continually gaining support for control system software.[17] Along with the comprehensive set of traditional IT vulnerabilities, it has recently included various control system vulnerabilities in its database. Nessus has also incorporated credential-based scanning capabilities that do not require network probing. While this feature significantly reduces the likelihood of impact system availability, it is only available on well-known operating systems.

Table 8.2 provides an overview of the introduced identification and analysis tools.

8.2.4 Target Vulnerability Validation

The vulnerability validation phase attempts to corroborate any previously determined vulnerability concerns. Validation plays a key role

Table 8.2   Identification and Analysis Tools

TOOL	TARGETED VULNERABILITIES	NEGATIVE IMPACT	DOMAIN SUPPORT
NMap	Network configurations and service/OS detection	High	Partial
Nessus	Operating system/services vulnerabilities and configurations	High	Partial

Note: OS = operating system.

Table 8.3   Vunerability Validation Tools

TOOL	TARGETED VULNERABILITIES	NEGATIVE IMPACT	DOMAIN SUPPORT
Metasploit	Vulnerability exploitation	High	Limited

within the power grid as vulnerabilities within many protocols and software platforms are not well known. Attempts to confirm the existence of a vulnerability may be required before investing resources in devising and deploying a mitigation strategy. Unfortunately, this step is generally extremely intrusive as attempts to exploit vulnerabilities often leave systems in unstable states. Activities in this phase should be performed on a replicated testing environment instead of critical operational systems. Some tools are available to assist with the vulnerability validation process. One example is the Metasploit Framework, an exploit development tool, which has recently gained some SCADA-specific capabilities to complement its expansive collection of traditional IT exploits (Table 8.3).[18]

8.2.5 Postexecution

The postexecution phase requires the evaluation of a vulnerability's potential system impacts and identification of mitigation techniques and any reporting responsibilities. While impact analysis has been addressed in IT systems through various quantitative and qualitative methods, these methods have not yet targeted a cyber physical system such as the smart grid. Determining impact within this domain may require additional research to detect the actual physical impact from a potential exploitation. Mitigation efforts also vary greatly with the grid. Often, software and field devices are not strongly supportive of upgrades and may require increased cost due to lack of remote accessibility. Therefore, various methods, such as network reconfigurations or increased detection capabilities, may be required to sufficiently address assessment findings.

8.3 State-of-Practice Review

The previous sections discussed the process of performing a vulnerability assessment tailored toward a substation automation environment. This section continues this state-of-practice review analysis by identifying current research efforts to provide improved capabilities within the domain. The process of identifying new vulnerabilities, improving detection within deployed systems, and managing them after their discovery presents many research challenges. Major efforts by industry and government are identified and then categorized based on their targeted impact. Table 8.4 provides a comprehensive review of these efforts.

8.4 Summary

The discovery of cyber vulnerabilities is becoming increasingly important within the smart grid due to an increased dependency on communication and computation for grid control. While assessment technologies and methodologies have been developed for the traditional computing environment, the transition to the substation automation environment is not well defined.

This chapter identified requirements for vulnerability assessments within smart grid environments, specifically identification of substation automation systems. A comprehensive methodology was introduced to identify the required steps within the process and detail how their application to this domain differs from traditional IT environments. Specific concerns were addressed, including the possibilities of negatively impacting the operational system through testing activities. Examples of security concerns were identified based on popular SCADA protocols and communication architectures. Finally, a review of current government and industry efforts within the vulnerability assessment domain was presented along with both current and future assessment tools.

Table 8.4 Vulnerability Management State of Practice

EFFORT	DESCRIPTION	TARGET
POLICY		
STANDARDS		
NIST 800-82[8]	Identification of vulnerabilities, network architecture models, and standards for security controls	ISC
NISTIR 7628[19]	Cybersecurity controls to address the increased connectivity within the smart grid	Smart grid
DHS CSET	Compliance/standards management and evaluation tool	SCADA
COMPLIANCE		
NERC CIP[3]	Enforceable vulnerability assessment requirements for bulk power systems	SCADA
NIST 800-53[20]	Enforceable security controls for government control system	ISC
DISCOVER		
DISCLOSURE		
NIST NVD[21]	Detailed database of known software vulnerabilities and misconfigurations	IT
ISC-CERT[22]	Publishes advisories on newly discovered vulnerabilities with control system software platforms	ISC
Vendor advisories	Vendor-released vulnerability information	ISC
TEST BEDS		
NSTB[23]	National laboratory collaboration with actual SCADA hardware/software for vulnerability assessment targeting without impact concerns	SCADA
Academic	For example, Iowa State University and University of Illinois,[24,25] realistic SCADA hardware/software, simulated power systems	SCADA
MANAGEMENT		
IMPACT ANALYSIS		
CVSS[26]	Non-ISC-specific scoring system for vulnerability criticality	IT
TESTING/DEPLOYMENT		
ISC-CERT	Mitigation recommendations based on vendor suggestions and ISC best practices	ISC

Note: CSET = cyber security evaluation tool, CVSS = common vulnerability scoring system, ISC-CERT = Industrial Control Systems Cyber Emergency Response Team, NISTIR = National Institute of Standards and Technology Interagency Report, NSTB = National SCADA Test Bed, DVD = National Vulnerability Database.

References

1. Government Accountability Office (GAO), *GAO-04-354: Critical Infrastructure Protection Challenges and Efforts to Secure Control Systems.* Washington, DC: U.S. GAO (March 2004).
2. Government Accountability Office (GAO), *GAO-05-434: Department of Homeland Security Faces Challenges in Fulfilling Cybersecurity Responsibilities.* Washington, DC: U.S. GAO (May 2005).
3. North American Electric Reliability Corporation (NERC), *NERC Critical Infrastructure Protection (CIP) Reliability Standards.* Atlanta, GA: NERC (2009).
4. K. Stouffer, J. Falco, and K. Scarfone, *NIST SP 800-115: Technical Guide to Information Security Testing and Assessment.* Gaithersburg, MD: National Institute of Standards and Technology (September 2008).
5. National Institute of Standards and Technology (NIST), *NIST SP 800-53A: Guide for Assessing the Security Controls in Federal Information Systems and Organizations, Building Effective Security Assessment Plans.* Gaithersburg, MD: NIST (June 2010).
6. Institute for Security and Open Methodologies (ISECOM), *Open Source Security Testing Methodology Manual (OSSTMM)* (2010). http://www.isecom.org/osstmm/.
7. R. C. Parks, *SAND2007-7328: Guide to Critical Infrastructure Protection Cyber Vulnerability Assessment.* Albuquerque, NM: Sandia National Laboratories (November 2007).
8. K. Stouffer, J. Falco, and K. Scarfone, *NIST SP 800-82: Guide to Industrial Control Systems (ICS) Security.* Albuquerque, NM: National Institute of Standards and Technology (September 2008).
9. Institute of Electrical and Electronics Engineers, *IEEE Standard for Electric Power Systems Communications, Distributed Network Protocol (DNP3), IEEE Std 1815-2010*, pp. 1–775 (1, 2010). IEEE, New York. doi: 10.1109/IEEESTD.2010.5518537.
10. M. Majdalawieh, F. Parisi-Presicce, and D. Wijesekera, DNPSec: Distributed Network Protocol version 3 (DNP3) security framework. In K. Elleithy, T. Sobh, A. Mahmood, M. Iskander, and M. Karim, eds., *Advances in Computer, Information, and Systems Sciences, and Engineering,* pp. 227–234. Springer, Dordrecht, the Netherlands (2006).
11. S. Bromberger, *DNS as a Covert Channel Within Protected Networks.* Clackamas, OR: National Electric Sector Cyber Security Organization (NESCO) (January 2011).
12. Bandolier. *Digital Bond, Inc.* http://www.digitalbond.com/wp-content/uploads/2008/mktg/Bandolier.pdf
13. D. M. Nicol, W. H. Sanders, M. Seri, and S. Singh. Experiences validating the access policy tool in industrial settings. In *Proceedings of the 2010 43rd Hawaii International Conference on System Sciences, HICSS '10*, pp. 1–8. IEEE Computer Society, Washington, DC (2010).
14. Wireshark. *Wireshark: A Network Protocol Analyzer.* http://www.wireshark.org

15. G. Rueff, C. Thuen, and J. Davidson. *Sophia Proof of Concept Report*, Idaho National Laboratory (March 2010).

16. Nmap. *Nmap Security Scanner*. http://nmap.org

17. Nessus. *Tenable Network Security*. http://www.nessus.org/nessus/.

18. Metasploit. *Metasploit Framework. Rapid7*. http://www.metasploit.com/.

19. National Institute for Standards and Technology (NIST), *NISTIR 7628: Guidelines for Smart Grid Cyber Security*. Gaithersburg, MD: NIST (August 2010).

20. National Institute for Standards and Technology (NIST), *NIST SP 800-53: Recommended Security Controls for Federal Information Systems and Organizations*. Gaithersburg, MD: NIST (August 2009).

21. National Institute for Standards and Technology (NIST), *National Vulnerability Database*. Gaithersburg, MD: National Institute of Standards and Technology (NIST). http://nvd.nist.gov/.

22. Industrial Control Systems Cyber Emergency Response Team (ISC-CERT). *Department of Homeland Security (DHS) Control Systems Security Program (CSSP)*. http://www.us-cert.gov/control_systems/ics-cert/.

23. Idaho National Laboratory (INL), *Common Cyber Security Vulnerabilities Observed in Control System Assessments by the INL NSTB Program*. Idaho Falls: INL (November 2008).

24. D. C. Bergman, D. Jin, D. M. Nicol, and T. Yardley, The virtual power system testbed and inter-testbed integration, *Second Workshop on Cyber Security Experimentation and Test*, Montreal, Canada (August 2009).

25. A. Hahn, B. Kregel, M. Govindarasu, J. Fitzpatrick, R. Adnan, S. Sridhar, and M. Higdon, Development of the PowerCyber SCADA security testbed. In *Proceedings of the Sixth Annual Workshop on Cyber Security and Information Intelligence Research, CSIIRW'10*, pp. 21:1–21:4. ACM, New York (2010).

26. K. Scarfone and P. Mell, An analysis of CVSS version 2 vulnerability scoring, *Third International Symposium on Empirical Software Engineering and Measurement*, October 15–16, 2009, Lake Buena Vista, FL (2009).

9

Smart Grid, Automation, and SCADA System Security

YONGGE WANG

Contents

In this chapter, we discuss the challenges for secure smart energy grid and automation systems. We first describe the current security status and existing attacks on power grid and critical infrastructures. Then, we use the supervisory control and data acquisition (SCADA) system as an example to show the challenges in securing the automation and smart power grid systems. Distributed control systems (DCSs) and SCADA systems were developed to reduce labor costs and to allow systemwide monitoring and remote control from a central location. Control systems are widely used in such critical infrastructures as the smart electric grid, natural gas, water, and wastewater industries. While control systems can be vulnerable to a variety of types of cyber attacks that could have devastating consequences, little research has been done to secure the control systems. The American Gas Association (AGA), International Electrotechnical Commission Technical

Committee Working Group 15 (IEC TC 57 WG 15), Institute of Electrical and Electronics Engineers (IEEE), National Institute of Standards and Technology (NIST), and National SCADA Test Bed Program have been actively designing cryptographic standards to protect SCADA systems. In this chapter, we briefly review these efforts and discuss related security issues.

9.1 Energy Grid and Supervisory Control and Data Acquisition: A High-Level Introduction

As stated in a Department of Energy (DoE) smart grid white paper,[1] the United States is in the process of modernization of the nation's electricity transmission and distribution system "to maintain a reliable and secure electricity infrastructure that can meet future demand growth" (Sec. 1301, p. 1). The major characterizations[1] of a modern electrical grid system include

- Improved reliability, security, and efficiency of energy distribution based on modern digital communication and control techniques
- Integration of industries involved in production and sale of energy, including the gas industry (e.g., natural gas extraction and distribution systems), the electrical power industry, the coal industry, and renewable resources (e.g., solar and wind power)
- Integration of demand response technologies such as real-time, automated, interactive technologies that optimize the physical operation of appliances and consumer devices for energy generation, transmission, distribution, and retailing (e.g., metering)
- Deployment of advanced electricity storage and peak-shaving technologies
- Availability of real-time information and control options to consumers
- Integration of cybersecurity techniques within the grid systems

In summary, the smart grid system is a secure and intelligent energy distribution system that delivers energy from suppliers to consumers based on two-way demand and response digital communication technologies to control appliances at consumers' homes to save energy and increase reliability. The smart grid system overlays the existing energy distribution system with digital information management and advanced metering systems. It is obvious that the increased interconnection and automation over the grid systems presents new challenges for deployment and management.

It is challenging to securely and efficiently convert the existing power grid systems to a smart system with these characteristics. According to the U.S. Energy Information Administration Web site,[2] at the end of 2010 there were more than 9,200 electric-generating plants in the United States, including coal, petroleum liquids, petroleum coke, natural gas, other gases, nuclear, hydroelectric, renewables, hydroelectric pumped storage, and other types. These generating plants produced 312,334,000 MWh of electricity during February 2011. The electricity is distributed to consumers via more than 300,000 miles of transmission lines throughout the United States. This power infrastructure was designed for performance rather than security, and the integrated communications protocols were designed for bandwidth efficiency without the consideration of cybersecurity. When moving the current energy distribution infrastructure toward a smart grid, we have to overcome the challenges of integrating network-based security solutions with automation systems, which usually requires a combination of new and legacy components and may not have enough reserved resources to perform security functionalities. In this chapter, we use supervisory control and data acquisition (SCADA) as an example to illustrate the strategies that may be employed for the design of smart grid systems.

Control systems are computer-based systems used within many critical infrastructures and industries (e.g., electric grid, natural gas, water, and wastewater industries) to monitor and control sensitive processes and physical functions. To deploy the smart grid system, there is a trend toward interconnecting SCADA systems and data networks (e.g., intranet). Thus, without a secure SCADA system it is impossible to deploy intelligent smart grid systems.

Typically, control systems collect sensor measurements and operational data from the field, process and display this information, and relay control commands to local or remote equipment. Control systems may perform additional control functions, such as operating railway switches and circuit breakers and adjusting valves to regulate flow in pipelines. The most sophisticated ones control devices and systems at an even higher level.

Control systems have been in place since the 1930s; there are two primary types of control systems: distributed control systems (DCS) and SCADA systems. DCS systems typically are used within a single processing or generating plant or over a small geographic area. SCADA systems typically are used for large, geographically dispersed distribution operations. For example, a utility company may use a DCS to generate power and a SCADA system to distribute it. We concentrate on SCADA systems, and our discussion is generally applicable to DCS systems.

9.2 Recent Attacks and Accidents with Energy Systems and Automation Systems

Several (real and simulated) attacks on energy and SCADA systems were reported in the past few years.[3–13] In the 2000 Maroochy Shire attack,[3] an Australian man hacked into the Maroochy Shire, Queensland, computerized waste management system and caused 200,000 gallons of raw sewage to spill out into local parks, rivers, and even the grounds of a Hyatt Regency hotel. It is reported that 49-year-old Vitek Boden had conducted a series of electronic attacks on the Maroochy Shire sewage control system after his job application had been rejected. Later investigations found radio transmitters and computer equipment in Boden's car. The laptop hard drive contained software for accessing and controlling the sewage SCADA systems.

By exploiting a vulnerability in a control system, the simulated Aurora generator test[5] conducted in March 2007 by the U.S. Department of Homeland Security resulted in a hacker's remote access to the generator room at the Idaho National Laboratory and the partial destruction of a $1-million diesel-electric generator.

In September 2007, an individual who claimed to be a CUPE (Canadian Union of Public Employees) member hacked into the city computer system in Vancouver that commands the town's traffic lights and set the computer clock 7 h behind.[6] The result was that traffic signals geared for midnight were managing traffic for the morning rush hour.

On April 8, 2009, an article[7] in the *Wall Street Journal* by Gorman reported that "cyberspies have penetrated the U.S. electrical grid and left behind software programs that could be used to disrupt the system, according to current and former national-security officials" (page 1). The same article mentioned that instead of damaging the power grid or other key infrastructures, the goals of these attacks were to navigate the U.S. electrical system and its controls to map them. To make things worse, these attacks were mainly detected by U.S. intelligence agencies instead of the companies in charge of the infrastructures. In other words, the U.S. utility companies are not ready for the protection of their current infrastructure, let alone the future interconnected smart grid systems. These attacks increase worries about cyber attackers who may take control of electrical facilities, a nuclear power plant, financial networks, or water, sewage, and other infrastructure systems via the Internet.

On Thursday, August 14, 2003, at approximately 4:11 p.m., a widespread power outage occurred throughout parts of the northeastern and midwestern United States and Ontario, Canada. According to a report by the New York Independent System Operator (NYISO),[8] this northeastern blackout of 2003 affected approximately 10 million people in Ontario and 45 million people in eight U.S. states; the NYISO megawatt load had a loss of 80% at the height of the outage. The final report[14] by the U.S.-Canada Power System Outage Task Force showed that the blackout was triggered by a race condition software bug in General Electric Energy's Unix-based XA/21 energy management system. The bug caused a disruption of service at FirstEnergy's control room, and the alarm system there stopped working for over an hour. After the alert system failure, neither audio nor visual alerts for important changes in system state were available to the operators. The unprocessed events queued up quickly, and the primary server failed within 30 minutes. Then, the server applications (including the failed alert systems) were automatically transferred to the backup server, which failed soon after. The lack of alarms led operators to dismiss a call from American Electric Power (AEP) about the tripping and reclosure of a 345-kV shared line in northeastern Ohio. FirstEnergy's technical support informed control room operators concerning the alarm system just before the massive blackout started.[15]

Although the software bug triggered this blackout, the U.S.-Canada Power System Outage Task Force report[14] listed four major causes for the blackout:

1. FirstEnergy (FE) and its reliability council "failed to assess and understand the inadequacies of FEs system, particularly with respect to voltage instability and the vulnerability of the Cleveland-Akron area, and FE did not operate its system with appropriate voltage criteria" (page 17).
2. FirstEnergy "did not recognize or understand the deteriorating condition of its system" (page 17).
3. FirstEnergy "failed to manage adequately tree growth in its transmission rights-of-way" (page 17).
4. There was "failure of the interconnected grids reliability organizations to provide effective real-time diagnostic support" (page 17).

The affected infrastructure of the blackout included power generation (power plants automatically went into "safe mode" to prevent damage in the case of an overload); water supply (some areas lost water pressure because pumps did not have power); transportation (trains had no power, and passenger security checking at affected airports ceased); communication systems (cellular communication devices were disrupted, radio stations were momentarily knocked off the air, and cable television systems were disabled); manufacturing (large numbers of factories were closed in the affected area, and freeway congestion in affected areas affected the "just-in-time" supply system).

In June 2010, it was reported[9,16] that the Stuxnet worm spreads around the world (with 59% infected systems in Iran) to subvert SCADA systems. Stuxnet malware targets only Siemens SCADA applications PCS 7, WinCC, and STEP7 that run on Microsoft Windows and Siemens S7 programmable logic controller (PLC). The worm initially spreads using USB (universal serial bus) flash drives and then uses four zero-day exploits to infect the Siemens SCADA and HMI (human-machine interface) system SIMATIC WinCC and PCS 7. Once infected, it attacks PLC systems with variable-frequency drives that spin between 807 and 1,210 Hz. When certain criteria are met, Stuxnet periodically modifies the frequency to 1,410 Hz, then to 2 Hz, and then to 1,064 Hz and thus affects the operation of the connected motors by changing their rotational speed.

In the 2009 Black Hat conference in Las Vegas, Nevada, Mike Davis[10] showed a simulation environment in which an attacker could take control of 15,000 of 22,000 home smart meters within 24 h by exploiting design flaws within an unnamed brand of smart meters.

Since November 2009, there have been reported[11] coordinated covert and targeted cyber attacks against global oil, energy, and petrochemical companies. These attacks are called the Night Dragon by McAfee.[11] An attack first compromises company extranet Web servers through Structured Query Language (SQL) injection techniques and then uploads some commonly available hacker tools to the compromised Web servers, which will allow the attacker to break into the company's intranet and obtain access to some sensitive internal desktops and servers. By disabling Microsoft Internet Explorer (IE) proxy settings, the attacker achieves direct communication from infected machines to the Internet. The attacker proceeds further to connect to other machines (targeting executives) and exfiltrating e-mail archives and other sensitive documents.

According to Zetter,[12] in May 2011, NSS Labs[17] researchers only spent 2 months on testing a few SCADA control systems and found several vulnerabilities in Siemens PLC and SCADA control systems that could be exploited by hackers to obtain remote access to the control systems to cause physical destruction to factories and power plants. It should be noted that Siemens PLC and SCADA systems are widely used in the world, controlling critical infrastructure systems such as nuclear power and enrichment plants and commercial manufacturing facilities. Under pressure by the Department of Homeland Security, the NSS Labs did not disclose details before Siemens could patch the vulnerabilities. This example shows that when the control systems are interconnected with the intranet, a dedicated attacker could easily mount serious attacks. It should also be noted that, in his dissertation, PhD student Sean Gorman from George Mason University using materials available publicly on the Internet (see, e.g., Blumenfeld[13] and Rappaport[18]), mapped every business and industrial sector in the American economy to the fiber-optic network that connects them. Similarly, under pressure from the government, Gorman's dissertation has never been made public.

9.3 SCADA Security

In this section, we demonstrate the challenges to secure the current automation systems, such as SCADA systems with examples. Part of these analysis were taken from the work of Wang.[19] In a typical SCADA system,[20] data acquisition and control are performed by remote terminal units (RTUs) and field devices that include functions for communications and signaling. SCADA systems normally use a poll response model for communications with clear text messages. Poll messages are typically small (less than 16 bytes), and responses might range from a short "I am here" to a dump of an entire day's data. Some SCADA systems may also allow for unsolicited reporting from remote units. The communications between the control center and remote sites could be classified into the following four categories.

1. *Data acquisition*: The control center sends poll (request) messages to RTUs, and the RTUs dump data to the control center. In particular, this includes *status scan* and *measured value scan*. The control center regularly sends a status scan request to remote sites to obtain field devices status (e.g., OPEN or CLOSED or a fast CLOSED-OPEN-CLOSED sequence) and a measured value scan request to obtain measured values of field devices. The measured values could be analog values or digitally coded values and are scaled into engineering format by the front-end processor (FEP) at the control center.
2. *Firmware download*: The control center sends firmware downloads to remote sites. In this case, the poll message is larger (e.g., larger than 64,000 bytes) than other cases.
3. *Control functions*: The control center sends control commands to an RTU at remote sites. Control functions are grouped into four subclasses: individual device control (e.g., to turn on/off a remote device); control messages to regulating equipment (e.g., a RAISE/LOWER command to adjust the remote valves); sequential control schemes (a series of correlated individual control commands); and automatic control schemes (e.g., closed control loops).
4. *Broadcast*: The control center may broadcast messages to multiple RTUs. For example, the control center broadcasts an emergent shutdown message or a set-the-clock-time message.

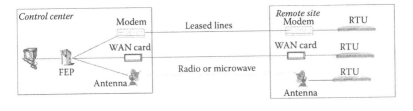

Figure 9.1   A simple SCADA system. WAN, wide-area network.

Acquired data are automatically monitored at the control center to ensure that measured and calculated values lie within permissible limits. The measured values are monitored with regard to rate of change and for continuous trend monitoring. They are also recorded for postfault analysis. Status indications are monitored at the control center with regard to changes and time tagged by the RTUs. In legacy SCADA systems, existing communication links between the control center and remote sites operate at very low speeds (could be on an order of 300 to 9,600 bps). Note that present deployments of SCADA systems have variant models and technologies, which may have much better performances (for example, 61850-based systems). Figure 9.1 describes a simple SCADA system.

In practice, more complicated SCADA system configurations exist. Figure 9.2 lists three typical SCADA system configurations (see, e.g., Report No. 12 of the American Gas Association [AGA][21]).

Recently, there have been several efforts to secure the national SCADA systems. Examples exist for the following companies and standards:

1. American Gas Association.[21] The AGA was among the first to design a cryptographic standard to protect SCADA systems. The AGA had originally been designing a cryptographic standard to protect SCADA communication links; the finished report is AGA 12, part 1. AGA 12, part 2, has been transferred to the Institute of Electrical and Electronics Engineers (IEEE) (IEEE 1711).

2. IEEE 1711.[22] This was transferred from AGA 12, part 2. This standard effort tries to define a security protocol, the Serial SCADA Protection Protocol (SSPP), for control system serial communication.

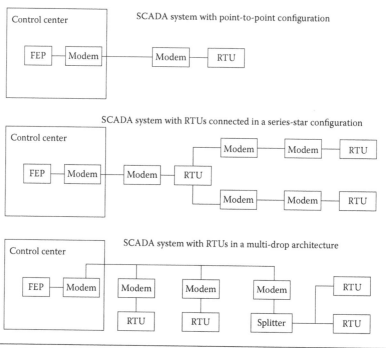

Figure 9.2 Typical SCADA system configurations.

3. IEEE 1815.[23] *Standard for Electric Power Systems Communications—Distributed Network Protocol (DNP3).* The purpose of this standard is to document and make available the specifications for the DNP3 protocol.

4. International Electrotechnical Commission Technical Committee Working Group 15 (IEC TC 57 WG 15).[24,25] The IEC TC 57 WG 57 standardized SCADA communication security via its IEC 608705 series.

5. National Institute of Standards and Technology (NIST).[26] The NIST Industrial Control System Security (ICS) group works on general security issues related to control systems such as SCADA systems.

6. National SCADA Test Bed Program.[27] The DoE established the National SCADA Test Bed program at Idaho National Laboratory and Sandia National Laboratory to ensure the secure, reliable, and efficient distribution of power.

9.3.1 *Threats to SCADA Systems*

SCADA systems were not designed with public access in mind; they typically lack even rudimentary security. However, with the advent of technology, particularly the Internet, much of the technical information required to penetrate these systems is widely discussed in the public forums of the affected industries. Critical security flaws for SCADA systems are well known to potential attackers. It is feared that SCADA systems can be taken over by hackers, criminals, or terrorists. Some companies may assume that they use leased lines and therefore nobody has access to their communications. The fact is that it is easy to tap these lines.[28] Similarly, frequency-hopping spread-spectrum radio and other wireless communication mechanisms frequently used to control RTUs can be compromised as well.

Several efforts[26,27,29] have been made for the analysis and protection of SCADA system security. According to these reports,[26,27,29] the factors that have contributed to the escalation of risk to SCADA systems include the following:

- The adoption of standardized technologies with known vulnerabilities. In the past, proprietary hardware, software, and network protocols made it difficult to understand how SCADA systems operated—and therefore how to hack into them. Today, standardized technologies such as Windows, Unix-like operating systems, and common Internet protocols are used by SCADA systems. Thus, the number of people with knowledge to wage attacks on SCADA systems has increased.
- The connectivity of control systems to other networks. To provide decision makers with access to real-time information and allow engineers to monitor and control the SCADA systems from different points on the enterprise networks, the SCADA systems are normally integrated into the enterprise networks. Enterprises are often connected to partners' networks and to the Internet. Some enterprises may also use wide-area networks and the Internet to transmit data to remote locations. This creates further security vulnerabilities in SCADA systems.

- Insecure remote connections. Enterprises often use leased lines, wide-area networks/Internet, and radio/microwave to transmit data between control centers and remote locations. These communication links could be easily hacked.
- The widespread availability of technical information about control systems. Public information about infrastructures and control systems is readily available to potential hackers and intruders. Sean Gorman's dissertation (see, e.g.,[13,18]), mentioned previously, is a good example for this scenario. Significant information on SCADA systems is publicly available (from maintenance documents, from former employees, and from support contractors, etc.). All these information sources could assist hackers in understanding the systems and finding ways to attack them.

Hackers may attack SCADA systems with one or more of the following actions:

1. Causing denial-of-service attacks by delaying or blocking the flow of information through control networks
2. Making unauthorized changes to programmed instructions in RTUs at remote sites, resulting in damage to equipment, premature shutdown of processes, or even disabling of control equipment.
3. Sending false information to control system operators to disguise unauthorized changes or to initiate inappropriate actions by system operators
4. Modifying the control system software, producing unpredictable results
5. Interfering with the operation of safety systems

The analysis in reports[26,27,29] showed that securing control systems poses significant challenges, which include

1. The limitations of current security technologies in securing control systems. Existing Internet security technologies such as authorization, authentication, and encryption require more bandwidth, processing power, and memory than control system components typically have. Controller stations are generally designed to do specific tasks, and they often use low-cost, resource-constrained microprocessors.

2. The perception that securing control systems may not be economically justifiable.

3. The conflicting priorities within organizations regarding the security of control systems. In this chapter, we concentrate on the protection of SCADA remote communication links. In particular, we discuss the challenges for protection of these links and design new security technologies to secure SCADA systems.

9.3.2 Securing SCADA Remote Connections

Relatively cheap attacks could be mounted on SCADA system communication links between the control center and RTUs since there is neither authentication nor encryption on these links. Under the umbrella of NIST's Critical Infrastructure Protection Cybersecurity of Industrial Control Systems, the AGA SCADA Encryption Committee has been trying to identify the functions and requirements for authenticating and encrypting SCADA communication links. Their proposal[21] is to build cryptographic modules that could be invisibly embedded into existing SCADA systems (in particular, one could attach these cryptographic modules to modems, such as those of Figure 9.2) so that all messages between modems are encrypted and authenticated when necessary, and they have identified the basic requirements for these cryptographic modules. However, due to the constraints of SCADA systems, no viable cryptographic protocols have been identified to meet these requirements. In particular, the challenges for building these devices are[21]

1. Encrypting of repetitive messages.
2. Minimizing delays due to cryptographic operations.
3. Ensuring integrity with minimal latency:
    • Intramessage integrity: If cryptographic modules buffer a message until the message authenticator is verified, it introduces message delays that are not acceptable in most cases.
    • Intermessage integrity: Reorder messages, replay messages, and destroy specific messages.
4. Accommodating various SCADA poll response and retry strategies: Delays introduced by cryptographic modules may

interfere with the SCADA system's error-handling mechanisms (e.g., time-out errors).

5. Supporting broadcast messages.
6. Incorporating key management.
7. Controlling the cost of devices and management.
8. Dealing with a mixed mode: Some SCADA systems have cryptographic capabilities; others do not.
9. Accommodating different SCADA protocols: SCADA devices are manufactured by different vendors with different proprietary protocols.

Wang[19] has recently designed efficient cryptographic mechanisms to address these challenges and to build cryptographic modules as recommended in AGA Report No. 12.[21] These mechanisms can be used to build plug-in devices called sSCADA (secure SCADA) devices that could be inserted into SCADA networks so that all communication links are authenticated and encrypted. In particular, authenticated broadcast protocols are designed so that they can be cheaply included into these devices. It has been a major challenging task to design efficiently authenticated emergency broadcast protocols in SCADA systems.

9.3.3 sSCADA Protocol Suite

The sSCADA protocol suite[19] is proposed to overcome the challenges discussed in the previous section. A sSCADA device installed at the control center is called a master sSCADA device, and sSCADA devices installed at remote sites are called slave sSCADA devices. Each master sSCADA device may communicate privately with several slave sSCADA devices. Occasionally, the master sSCADA device may also broadcast authenticated messages to several slave sSCADA devices (e.g., an emergency shutdown). An illustrative sSCADA device deployment for point-to-point SCADA configuration is shown in Figure 9.3.

It should be noted that the AGA had originally designed a protocol suite to secure the SCADA systems[21,30] (an open source implementation could be found in Reference 31). However, Wang[19] has broken these protocol suites by mounting a replay attack.

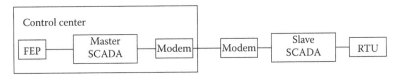

Figure 9.3 sSCADA with point-to-point SCADA configuration.

To reduce the cost of sSCADA devices and management, only symmetric key cryptographic techniques are used in our design. Indeed, due to the slow operations of public key cryptography, public key cryptographic protocols could introduce delays in message transmission that are not acceptable to SCADA protocols. Semantic security property[32] is used to ensure that an eavesdropper has no information about the plaintext, even if the eavesdropper sees multiple encryptions of the same plaintext. For example, even if the attacker has observed the ciphertexts of "shut down" and "turn on," it will not help the attacker to distinguish whether a new ciphertext is the encryption of "shut down" or "turn on." In practice, the randomization technique is used to achieve this goal. For example, the message sender may prepend a random string (e.g., 128 bits for Advanced Encryption Standard [AES] 128) to the message and use special encryption modes such as chaining block cipher (CBC) mode or hash-CBC (HCBC) mode. In some modes, this random string is called the initialization vector (IV). This prevents information leakage from the ciphertext even if the attacker knows several plaintext/ciphertext pairs encrypted with the same key.

Since SCADA communication links could be as low as 300 bps and immediate responses are generally required, there is no sufficient bandwidth to send the random string (IV) each time with the ciphertext; thus, we need to design different cryptographic mechanisms to achieve semantic security without additional transmission overhead. In our design, we use two counters shared between two communicating partners, one for each direction of communication.

The counters are initially set to zeros and should be at least 128 bits, which ensures that the counter values will never repeat, avoiding replay attacks. The counter is used as the IV in message encryptions if CBC or HCBC mode is used. After each message encryption, the counter is increased by one if CBC mode is used, and it is increased by the number of blocks of encrypted data if the HCBC mode is used. The two communicating partners are assumed to know the values of the counters, and the counters do not need to be added to each ciphertext. Messages may become lost, and the two counters need to be synchronized occasionally (e.g., at off-peak time). A simple counter synchronization protocol is proposed for the sSCADA protocol suite. The counter synchronization protocol could also be initiated when some encryption/decryption errors appear due to unsynchronized counters.

For two sSCADA devices to establish a secure channel, a master secret key needs to be bootstrapped into the two devices at deployment time (or when a new sSCADA device is deployed into the existing network). For most configurations, secure channels are needed only between a master sSCADA device and a slave sSCADA device. For some configurations, secure channels among slave sSCADA devices may also be needed. The secure channel identified with this master secret is used to establish other channels, such as session secure channels, time synchronization channels, authenticated broadcast channels, and authenticated emergency channels.

Assume that $\mathcal{H}(\cdot)$ is a pseudorandom function (e.g., constructed from Secure Hash Algorithm [SHA]-256) and two sSCADA devices A and B share a secret $\mathcal{K}_{AB} = \mathcal{K}_{BA}$. Depending on the security policy, this key $\mathcal{K}_{AB}$ could be the shared master secret or a shared secret for one session that could be established from the shared master key using a simple key establishment protocol (to achieve session key freshness, typically one node sends a random nonce to the other one, and the other node sends the encrypted session key together with an authenticator on the ciphertext and the random nonce). Keys for different purposes could be derived from this secret as follows (it is not a good practice to use the same key for different purposes): For example, $K_{AB} = \mathcal{H}(\mathcal{K}_{AB}, 1)$ is for message encryption from A to B, $K'_{AB} = \mathcal{H}(\mathcal{K}_{AB}, 2)$ is for message authentication from A to B, $K_{BA} = \mathcal{H}(\mathcal{K}_{AB}, 3)$ is for message encryption from B to A, and $K'_{BA} = \mathcal{H}(\mathcal{K}_{AB}, 4)$ is for message authentication from B to A.

Optional message authentication codes (MACs) are used for two parties to achieve data authentication and integrity. MACs that could be used for sSCADA implementation include HMAC,[33,34] CBC-MAC,[35] and others. When party A wants to send a message m to party B securely, A computes the ciphertext $c = \mathcal{E}(C_A, K_{AB}, \overline{c_A} \| m)$ and message authenticator $mac = MAC(K'_{AB}, C_A \| c)$, where $\overline{c_A}$ is the last l bits of $\mathcal{H}(C_A)$ (l could be as large as possible if bandwidth is allowed, and 32 bits should be the minimal), $\mathcal{E}(C_A, K_{AB}, \overline{c_A} \| m)$ denotes the encryption of $\overline{c_A} \| m$ using key K_{AB} and random-prefix (or IV) C_A, and C_A is the counter value for the communication from A to B. Then, A sends the following packets to B:

$$A \rightarrow B: \quad c, mac \text{ (optional)}$$

When B receives these packets, B decrypts c, checks that $\overline{c_A}$ is correct, and verifies the message authenticator mac if mac is present. As soon as B receives the first block of the ciphertext, B can check whether $\overline{c_A}$ is correct. If it is correct, then B continues the decryption and updates its counter. Otherwise, B discards the entire ciphertext. If the message authenticator code mac is present, B also verifies the correctness of mac. If mac is correct, B does nothing; otherwise, B may choose to inform A that the message was corrupted or try to resynchronize the counters.

There are several implementation issues on how to deliver the message to the target (e.g., RTU). For example, there are the following:

1. B uses the counter to decrypt the first block of the ciphertext; if the first l bits of the decrypted plaintext are not consistent with $\mathcal{H}(C_A)$, then the reason could be that the counter C_A is not synchronized or that the ciphertext is corrupted. B may try several possible counters until the counter-checking process succeeds. B then uses the verified counter and the corresponding key to decrypt the message and deliver each block of the resulting message to the target as soon as it is available. If no counter could be verified in a limited number of trials, B may notify A of the transmission failure and initiate the counter synchronization protocol in the next section. The advantage of this implementation is that we have minimized delay from the cryptographic devices, thus minimizing the interference of SCADA protocols. Note that in this implementation, the message authenticator mac is not used. If the ciphertext was tampered, we rely on the error correction mechanisms (normally CRC codes) in SCADA systems to discard the entire message. If CBC (respectively

HCBC) mode is used, then the provable security properties (respectively provable online cipher security properties) of CBC mode (respectively HCBC mode)[36,37] guarantee that the attacker has no chance to tamper with the ciphertext, so that the decrypted plaintext contains a correct CRC that was used by SCADA protocols to achieve integrity.

2. Proceed as in case 1. In addition, the *mac* is further checked, and the decrypted message is delivered to the SCADA system only if the *mac* verification passes. The disadvantage for this implementation is that these cryptographic operations introduce significant delay for message delivery, and it may interfere with SCADA protocols.

3. Proceed as in case 1. The decrypted message is delivered to the SCADA system as soon as available. After receiving the entire message and *mac*, B will also verify *mac*. If the verification passes, B will do nothing. Otherwise, B resynchronizes the counter with A or initiates some other exception-handling protocols.

4. To avoid delays introduced by cryptographic operations and to check the *mac* at the same time, sSCADA devices may deliver decrypted bytes immediately to the target except the last byte. If the message authenticator *mac* is verified successfully, the sSCADA device delivers the last byte to the target; otherwise, the sSCADA device discards the last byte or sends a random byte to the target. That is, we rely on the error correction mechanisms at the target to discard the entire message. Similar mechanisms have been proposed.[21] However, an attacker may insert garbage between the ciphertext and *mac*, thus tricking the sSCADA device to deliver the decrypted messages to the SCADA system. If this happens, we essentially do not receive an advantage from this implementation. Thus, this implementation is not recommended.

5. Instead of prepending $\overline{c_A}$ to the plaintext message, one may choose to prepend three bytes of other specially formatted string to the plaintext message (bandwidth of three bytes is normally available in SCADA systems) before encryption. This is an acceptable solution although we still prefer our solution of prepending the hash outputs of the counter.

There could be other implementations to improve the performance and interoperability with SCADA protocols. sSCADA devices should provide several possible implementations for users to configure. Indeed, sSCADA devices may also be configured in a dynamic way so that for different messages it uses different implementations.

In some SCADA communications, message authentication only is sufficient. That is, it is sufficient for A to send (m, mac) to B, where m is the cleartext message and $mac = MAC(K'_{AB}, C_A \| m)$. sSCADA devices should provide configuration options to perform message authentication without encryption. In this case, even if the counter value is not used as the IV, the counter value should still be authenticated in the mac and be increased after the operation. This will provide message freshness assurance and avoid replay attacks. sSCADA should also support message pass-through mode. That is, the message is delivered without encryption and authentication. In summary, it should be possible to configure an sSCADA device in such a way that some messages are authenticated and encrypted, some messages are authenticated only, and some messages are passed through directly.

9.3.4 Counter Synchronization

In the point-to-point message authentication and encryption protocol, we assume that both sSCADA devices A and B know each other's counter values C_A and C_B, respectively. In most cases, reliable communication in SCADA systems is provided, and the security protocols in the previous section work fine. Still, we provide a counter synchronization protocol so that sSCADA devices can synchronize their counters when necessary. The counter synchronization protocol could be initiated by either side. Assume that A initiates the counter synchronization protocol. Then, the protocol looks as follows:

$$A \rightarrow B: \quad N_A$$
$$B \rightarrow A: \quad C_B, MAC(K'_{BA}, N_A \| C_B)$$

The initial counter values of two sSCADA devices could be bootstrapped directly. The counter synchronization protocol presented could also be used by two devices to bootstrap the initial counter values. A master sSCADA device may also use the authenticated broadcast channel that we discuss in the next section to set the counters of several slave sSCADA devices to the same value using one message.

9.4 Conclusion

In this chapter, we discussed the challenges for smart grid system security. We then use control systems (in particular, SCADA systems) as examples for studying how to address these challenges. In particular, we mentioned Wang's attack[19] on the protocols in the first version of the AGA standard draft.[30] This attack showed that the security mechanisms in the first draft of the AGA standard protocol could be easily defeated. We then proposed a suite of security protocols optimized for SCADA/DCS systems. These protocols are designed to address the specific challenges of SCADA systems.

Recently, there has been a wide interest in the secure design and implementation of smart grid systems.[38] The SCADA system is one of the most important legacy systems of the smart grid systems. Together with other efforts such as those offered in IEEE 1711,[22] IEEE 1815,[23] IEC TC 57,[24] IEC 60870-5,[25] NIST Industrial Control System Security,[26] and the National SCADA Testbed Program,[27] the work in this chapter presents an initial step for securing the SCADA section of the smart grid systems against cyber attacks.

References

1. Department of Energy. *Title XIII—Smart Grid* (2010). http://www. oe.energy.gov/DocumentsandMedia/EISA_Title_XIII_Smart_Grid. pdf
2. U.S. Energy Information Administration. *Net Generation by Energy Source: Total (All Sectors)* (2011). http://www.eia.gov/cneaf/electricity/epm/table1_1.html
3. M. Abrams and J. Weiss. *Malicious Control System Cyber Security Attack Case Study—Maroochy Water Services, Australia* (2010). http://csrc.nist. gov/groups/SMA/fisma/ics/documents/Maroochy-Water-Services-Case-Study_briefing.pdf
4. M. Abrams and J. Weiss. *Bellingham, Washington, Control System Cyber Security Case Study* (2007). http://csrc.nist.gov/groups/SMA/fisma/ics/documents/Bellingham_Case_Study_report2020Sep071.pdf

5. *USA Today*. AURORA case: U.S. video shows hacker hit on power grid (2007). http://www.usatoday.com/tech/news/computersecurity/2007-09-27-hacker-video_N.htm

6. SPAMfighter. Vancouver city-police investigating possible sabotage of traffic light computer system (2007). http://www.spamfighter.com/News_Show_Other.asp?M=10&Y=2007

7. S. Gorman. Electricity grid in us penetrated by spies. *Wall Street Journal* (April 8, 2009). http://online.wsj.com/article/SB123914805204099085.html

8. ISO New York Independent System Operator. *NYISO Interim Report on the August 14, 2003 Blackout* (2004). http://www.hks.harvard.edu/hepg/Papers/NYISO.blackout.report.8.Jan.04.pdf

9. G. Keizer. Is Stuxnet the "best" malware ever? (2010). http://www.infoworld.com/print/137598

10. M. Davis. Smartgrid device security adventures in a new medium (2009). http://www.blackhat.com/presentations/bh-usa-09/MDAVIS/BHUSA09-Davis-AMI-SLIDES.pdf

11. McAfee. Global energy cyberattacks: Night dragon (February 2011). http://www.mcafee.com/us/resources/white-papers/wp-global-energy-cyberattacks-night-dragon.pdf

12. K. Zetter. Fearing industrial destruction, researcher delays disclosure of new Siemens SCADA holes (2011). http://www.wired.com/threatlevel/2011/05/siemens-scada-vulnerabilities/.

13. L. Blumenfeld. Dissertation could be security threat. *Washington Post* (July 7, 2003). http://www.washingtonpost.com/ac2/wp-dyn/A23689-2003Jul7

14. U.S.-Canada Power System Outage Task Force. *Final Report on the August 14, 2003 Blackout in the United States and Canada: Causes and Recommendations* (April 2004). https://reports.energy.gov/BlackoutFinal-Web.pdf

15. North American Electric Reliability Council. *Technical Analysis of the August 14, 2003, Blackout: What Happened, Why, and What Did We Learn?* (2004). http://www.nerc.com/docs/docs/blackout/NERC_Final_Blackout_Report_07_13_04.pdf

16. N. Falliere, L. Murchu, and E. Chien. W32.stuxnet dossier (February 2011). http://www.symantec.com/content/en/us/enterprise/media/security_response/whitepapers/w32_stuxnet_dossier.pdf

17. NSS Labs. Home page. http://www.nsslabs.com/.

18. J. Rappaport. What you don't know might hurt you: alum's work balances national security and information sharing. http://gazette.gmu.edu/articles/11144

19. Y. Wang. sSCADA: Securing SCADA infrastructure communications, *International Journal Communication Networks and Distributed Systems* 6(1), 59–78 (2011).

20. T. Cegrell. *Power System Control Technology*. Prentice-Hall International, Harlow, UK (1986).

21. American Gas Association. *AGA Report No. 12. Cryptographic Protection of SCADA Communications: General Recommendations*. Draft 2, February 5, 2004. Draft 2 is no longer available online. Draft 3 (2010) is available for purchase. http://www.aga.org/.

22. Institute of Electrical and Electronics Engineers. *IEEE 1711. Trial Use Standard for a Cryptographic Protocol for Cyber Security of Substation Serial Links* (2011). http://standards.ieee.org/findstds/standard/1711-2010.html

23. Institute of Electrical and Electronics Engineers. *IEEE 1815. Standard for Electric Power Systems Communications—Distributed Network Protocol (DNP3)* (2010). http://grouper.ieee.org/groups/1815/.

24. International Electrotechnical Commission. *IEC TC 57. Focus on the IEC TC 57 Standards* (2010). http://www.ieee.org/portal/cms_docs_pes/pes/subpages/publications-folder/TC_57_Column.pdf

25. International Electrotechnical Commission. *IEC 60870-5. Group Maillist Information* (2010). http://www.trianglemicroworks.com/iec60870-5/index.htm

26. National Institute of Standards and Technology (NIST). *NIST Industrial Control System Security (ICS)* (2011). http://csrc.nist.gov/groups/SMA/fisma/ics/index.html

27. Idaho National Laboratory. National SCADA Testbed Program (2011). http://www.inl.gov/scada/.

28. Granite Island Group. Wiretapping and outside plant security—wiretapping 101 (2011). http://www.tscm.com/outsideplant.html

29. General Accounting Office. *GAO-04-628. Critical Infrastructure Protection: Challenges and Efforts to Secure Control Systems. Testimony Before the Subcommittee on Technology Information Policy, Intergovernmental Relations and the Census, House Committee on Government Reform* (March 30, 2004). http://www.gao.gov/new.items/d04628t.pdf

30. A. K. Wright, J. A. Kinast, and J. McCarty. *Low-Latency Cryptographic Protection for SCADA Communications*, In *Proc. 2nd Int. Conf. on Applied Cryptography and Network Security, ACNS 2004*, vol. 3809, *LNCS*, pp. 263–277. Springer-Verlag, New York (2004).

31. A. Wright. Scadasafe (2006). http://scadasafe.sourceforge.net

32. S. Goldwasser and S. Michali. Probabilistic encryption, *Journal of Computer and System Sciences* 28, 270–299 (1984).

33. M. Bellare, R. Canetti, and H. Krawczyk. Message authentication using hash functions—the HMAC construction, *RSA Laboratories CryptoBytes* 2(1) (Spring 1996).

34. H. Krawczyk, M. Bellare, and R. Canetti. *HMAC: Keyed-Hashing for Message Authentication*, Internet RFC 2104 (February 1997). http://www.itl.nist.gov/fipspubs/fip81.htm

35. National Institute of Standards and Technology (NIST). *DES Model of Operation*, FIPS Publication 81. NIST, Gaithersburg, MD (1981).

36. M. Bellare, A. Boldyreva, L. Knudsen, and C. Namprempre. On-line ciphers and the Hash-CBC constructions. In *Advances in Cryptology—Crypto 2001*, vol. 2139, *LNCS*, pp. 292–309. Springer Verlag, New York (2001).

37. M. Bellare, J. Kilian, and P. Rogaway. The security of the cipher block chaining message authentication code, *Journal of Computer and System Sciences* 6(3), 362–399 (2000).
38. Department of Energy. *Study of Security Attributes of Smart Grid Systems— Current Cyber Security Issues* (April 2009). http://www.inl.gov/scada/publications/d/securing_the_smart_grid_current_issues.pdf

10

SMART GRID SECURITY IN THE LAST MILE

TAE OH, SUMITA MISHRA, AND CLARK HOCHGRAF

Contents

Maintaining integrity, availability, authenticity, and confidentiality of smart grid data and control information becomes increasingly challenging in the last mile to the home. Physical security is more difficult to achieve, leading to greater potential for tampering and compromise of nodes. The larger number of nodes and the interdependency of nodes for communication leave the system more vulnerable to certain types of attacks. Encryption is essential to network security; however, encryption key management is a particular challenge in the smart grid due to the large number of distributed nodes. Interoperability and flexibility goals can appear to be at odds with the implementation of security measures that ensure valid data are being provided. A balance must be struck between competing objectives. Security decisions in the last mile must be evaluated from the broader perspective of maintaining operation of the smart power system in the face of evolving attacks and adversaries over the deployment life of the system.

10.1 Introduction

Collection of data and control of devices are two main objectives of the smart grid. Collected data may include energy usage, power consumption, local voltage, volt-ampere reactive (VAR) power, and operational status for numerous devices at numerous locations. Device control may include changing taps on transformers, engaging VAR compensation capacitors, disconnecting loads, reducing loads, delaying the start of a load, or changing command set points for distributed generators.

All of these tasks require communication of either information or control signals. If invalid, inaccurate, malicious, or untimely information is provided, the effects on the power system operation can be severe, including over- or undervoltage, interruption of power both locally and regionally, damage to connected equipment, hazards to personnel, and financial losses. The information gathered can be used for inappropriate, unexpected, or unsavory purposes by authorized and unauthorized third parties. As a result, both security and privacy of sensor data and control information are essential.

10.2 Smart Grid System Architecture in the Last Mile

A number of organizations, including the National Institute of Standards and Technology (NIST), Institute of Electrical and Electronics Engineers (IEEE), Electric Power Research Institute (EPRI), and others, have created architectural models of the smart grid. In the last mile, the smart grid includes power distribution equipment and over-lapping communication networks, for example, the field-area network (FAN), neighborhood-area network (NAN), automated metering infrastructure (AMI), and home-area networks (HANs).

Much of the emphasis in the smart grid is on smart meters (AMI) and linking AMI into other networks in the home for load control or sending pricing signals to homeowners, and in the power system for connecting AMI infrastructure to higher-level centralizing networks.

In Germany, a smart grid architecture with different slightly different terminology is used.[1] AMI smart metering devices are part of a local metrological network (LMN) that connects to the consumer's HAN and can control loads or generation units that are part of a controllable local system (CLS). The AMI/LMN network data is passed back toward a central data collector through a gateway that connects to a wide-area network (WAN).

The architectural description differences as well as different privacy standards influence how security and privacy solutions are achieved. For example, in Germany, the gateway acts as a firewall to the HAN and CLS from the WAN.

10.3 Control System Perspective: Impact of the Smart Grid on Electric Power System Stability

In the smart grid, end-consumer sensor data and controllable systems are integrated and aggregated into larger values of power consumption and larger effective controllable load systems. The data for aggregate power may be used for decisions at the level of a feeder, substation, or even region. Besides data aggregation, controlled local systems are aggregated and may be commanded to act as a large single controllable load. With the new controllability arising from the smart grid, it is useful to take a control systems perspective on the functioning of the smart grid system.

An adversary who manipulates either the load data or controllable local systems may be able to have a broad impact on the electric power system. By manipulating loads on and off, the adversary could reduce damping in the electric power system, modifying the system eigenvalues and worsening any latent stability issues. Combining such an attack with obfuscation of the system state by manipulation of sensor data may lead to a loss of system control.

Positive control benefits may also be achieved by adding active damping using aggregated controllable local systems. From a control system perspective, lag and latency become a critical issue in achieving a stable feedback control system using the smart grid controllable loads. Any feedback system with sufficient lag and gain can be made unstable. At a minimum, lag in feedback signals creates a reduction in stability. The lag or latency introduced by a particular security implementation must be considered in stability analyses.

10.4 Planning for Security and Privacy

Designing a smart grid system for security and privacy can be approached starting with a risk assessment that examines adversaries, their objectives, and threats. Adversaries may be insiders or outsiders. They may have physical access or they may not. Their objectives may be financial, political, or disruptive. Guidelines for assessing cyber-security risks are available from the National Association of State Energy Officials (NASEO)[2] and NIST.[3] The Utility Communication Architecture International User Group (UCAlug) has outlined secu-rity strategies and threats specific to AMI.[4]

A general strategy of applying security in layers and at different levels is recommended. This "defense-in-depth" strategy includes physical access control "fences and gates," role limitations, security logs, encryption, secure communication, and auditing of information-handling procedures and practices. Ensuring security and privacy goes beyond encryption and secure communication.

Challenges in smart grid security design include

1. Knowing who to trust (authentication)
2. Detecting intrusion, even if there is no disruption
3. Understanding how a potential attack affects system operation
4. Maintaining secure data-handling procedures (privacy) across organizations outside the utility (e.g., third parties, outsourced services)

Germany's security strategy for the AMI gateway[1] addresses chal-lenge 3 by specifically ensuring that if the communication network is disrupted, the system fail-safe is to ensure that electricity is still pro-vided to the consumer, with no possibility of impact on the delivery of the commodity.

Detecting intrusion, challenge 2, is addressed by security logs and tamper detection that is observable by both the consumer and the gate-way operator. The AMI gateways are also required to be installed in a nonpublic environment to reduce the potential for physical access to the equipment. Messages on the system have timestamps to prevent a replay attack, by which a copy of an authentic message is replayed at a later time.

In preventing an adversary from affecting the system, it is impor-tant that the adversary does not have access to the complete system image. Concealment of the network nodes, communication pathways, and power system architecture is recommended. This prevents the attacker from gaining relevant information by observing responses to a failed message attempt or from observing information flow.

Maintaining privacy of personal usage information can be accomplished by allowing upstream parties to have access to only the minimum amount of data needed for billing or system operation. Data should be encrypted and pseudonymized or aggregated as appropriate to anonymize it.

10.5 Security Threats in the Field-Area Network/ Neighbor-Area Network

10.5.1 Physical-Layer Attacks

Smart grids are expected to have nodes that are installed in areas considered to be outside premises. In these locations, they become highly vulnerable to physical damage due to environmental reasons or manhandling of nodes. Such damage poses a threat to the integrity of the entire smart grid network. At the physical (PHY) layer, it is also susceptible to break down of transmission medium and rogue capturing of nodes.[5] Such threats can be combated through the utilization of tamper-resistant and damage-impervious devices capable of sending security alerts. The use of encryption in devices and deployment of devices that securely store cryptographic keys and execute an authentication check on each link setup should be undertaken to thwart security threats at a physical level.

10.5.2 Link-Layer Attacks

In a NAN, nodes are allowed to join and leave dynamically, which leads to issues of securely communicating multicast messages at the link layer. Jamming of the communication medium due to reprobate capture of the network is another issue that needs consideration. Jamming can also occur in fast-hop ad hoc networks if the number of hops exceeds 1,000 hops/s, causing internal interference.[6] Most NAN networks are ad hoc networks having a medium access control (MAC) protocol responsible for allocating the medium and the available resources in a distributed manner. This makes the network susceptible to availability attacks by selfish nodes that monopolize the available resources. Radio-frequency (RF) spectrum jamming can be avoided using frequency-hopping spread spectrum (FHSS), which varies the channel from 50 to 100 times per second, making it difficult to lock on to one particular frequency.

With multiple types of traffic being carried on a converged smart grid network, quality of service (QoS) is important to ensure that critical control traffic is not delayed by less-critical traffic. Allowable latency times may be less than 3 ms for protection and safety-critical control communications. Some aspects may tolerate up to 160 ms of latency. Noncritical communications can handle latencies of greater than 160 ms. QoS requirements not only lead to multiple levels of security but also place a limit on the maximum tolerable processing time of security measures implemented at this layer.

One security concern pertaining characteristically to a smart grid network is sleep derivation or torture attacks. Almost all components in a smart grid are designed to have a long sleep time when the device is in the off state, which translates to breakdown of the device if an attack overloads it.

The MAC layer attacks for the smart grid are generally averted by security protocols that involve techniques like MAC ID filtering, QoS provisioning, and so on. Most PHY layer security measures can also be extended to the MAC layer.

10.5.3 Network-Layer Attacks

Network-layer attacks are generally characterized by attacks on routing tables, which affect data traffic flows. The routing table is responsible for relaying the messages to their correct destination. Network-layer attacks aim at modifying the routing table so that traffic flows through a specific node controlled by the attacker. The attacker then can generate messages with false information or erroneously relay information that may cause congestion in the network. Denial-of-service (DoS) attacks at the network layer can be undertaken by fabricating routing tables aimed at disrupting traffic flow and eavesdropping on the information transmitted in the smart grid.[7]

Network-layer attacks include the following:

- Routing black holes: A node is hacked and is then broadcast as the shortest path, resulting in all traffic being directed to the hacked node.

- Sybil attack: Some sensor nodes in the network are misguided into believing that nodes that either are multiple hops away or that do not exist are their neighbors.
- Wormholes: A considerable amount of the network traffic is tunneled from one place in the network to another distant place in the network, depriving other parts of the network.

Attacks can occur on neighbor-sensing protocols by inserting unauthorized nodes, which can be prevented through the routine use of encryption, integrity examination, and authentication mechanisms. However, this leads to an added security threat of attack to exploit route maintenance procedures.[8]

10.5.4 Internet Protocol Addressing Specific Attacks

The use of Internet Protocol (IP) addressing in smart grid communication does lead to confidentiality and authorization issues. IP spoofing, dual-stack convergence, and cyber attacks at this level are some other concerns. Cyber threats include cyber spies mapping the grid and installing malicious software capable of destroying or disrupting services. IP-based requirements were written for computers (hosts) and routers; some AMI nodes do not meet them and probably cannot meet them without further specification development. On the other hand, security measures for IP-based routing are well defined and can be added to the smart grid with some minor modifications.

10.5.5 Transport-Layer Attacks

The communications module inside each meter is connected to the meter via a serial port, which can be disconnected so that the meter does not report usage. Deploying smart meters, which are capable of detecting such disconnects and other types of tampering and reporting such incidents to operators, can mitigate service theft via meter/communications module interface intrusion. The primary transport protocols like TCP (Transmission Control Protocol), UDP (User Datagram Protocol), DCCP (Datagram Congestion Control Protocol), and SCTP (Stream Control Transmission Protocol) provide multiplexing of different traffic flows between two hosts, and the logical separation provided by the transport layer is not intended to guard against malicious attacks by a determined adversary.

10.5.6 Application-Layer Attacks

Attacks on the application pose a threat since cryptography and encryption are not enough to prevent them. Verifying the data received with statistical data corresponding to the model can prevent the attacks. While this is not a foolproof method, the method is useful.

10.5.7 Other Prominent Security Threats

10.5.7.1 Back-Office Compromise Back-office compromise could take place when individuals illegally gain access to the smart grid management database. From there, they could compromise the reliability of the entire grid, including unsanctioned access to billing and other back-office systems. This could lead to embezzlement of service in addition to loss of customer confidentiality.

By the same token, with access to the database that stores privileged data, an attacker could modify the credentials to which coordinators respond and potentially bring down the grid. Physical security, strong validation, authorization using multilevel privileges, and network access regulation using firewalls are all mechanisms that can be used to combat a back-office attack. Encryption of databases, password, and customer information should be undertaken, and right of entry to the control system should be restricted to specific physically safeguarded sites.

10.5.7.2 Grid Volatility The smart grid network has much intelligence at its edges, that is, at the entry point and at the end-user's meter and at the back office where all the data are accumulated and processed. However, in the grid network itself, there is insufficient intelligence governing the switching functions. This lack of integrated development makes the grid a volatile network with little if any software intelligence to control it, making the smart grid vulnerable to physical and cyber attacks in the middle.[9]

10.5.7.3 Security Discrepancy With the smart grid, there are multiple stakeholders with different agendas motivating them. Security standards have to be unbiased and account for security of the smart grid in its totality. Also, encryption and other security measures have to be maintained throughout the entire network as the network is only as strong as its weakest link.

10.6 Security of AMI System: AMI Issues and Current Weaknesses

Advanced metering infrastructure security is one of the key components in the smart grid infrastructure. There is a proposed AMI security specification under development that will provide the utility industry along with supporting vendor communities and other stakeholders a set of security requirements. The requirements should be applied to AMI implementations to ensure the high level of information assurance, availability and security necessary to maintain a reliable system and consumer confidence.[10]

10.6.1 AMI Components

The AMI system consists of several components interconnected to form a network architecture, which provides communication capabilities in a power grid. Some of the prominent components are as follows:[11]

Smart Meter: This meter provides energy-related information as well as metrological data. In addition, the meter provides periodic data for customer energy usage.

Customer Gateway: This gateway is an interface between the AMI network and HAN or building management system (BMS). The gateway location may be different from that of the smart meter.

AMI Communications Network: The network provides information flow from the smart meter to the AMI headend.

AMI Headend: The component provides a management function for information flow between an external system and the AMI network.

10.6.2 Security Issues in AMI Components

10.6.2.1 Confidentiality in an AMI System   The main issue of confidentiality in AMI systems[10,12] is privacy since customers would not want private firms, marketing agencies, or unauthorized people to access their energy or electric utilization patterns. Therefore, the concerned authority has to make sure that data leaks do not occur either intentionally or unintentionally.

The AMI communications network must also restrict unauthorized access or information passing between customers. It is important to keep energy and other information from the smart meter confidential even from physical tampering to access the stored data. Lack of confidentiality could result in a hacker accessing data that reveal which houses in an area are empty or tricking the meter to unwittingly pay for your neighbors' electricity. Also, the hacker could hijack the control of your energy usage, such as turning on or off smart grid-enabled household appliances. If the AMI system interfaced to the customer gateway into the HAN, a commercial energy management system, or other automated system, the privacy of those systems must be considered and maintained.

10.6.2.2 Integrity in AMI Systems   In AMI systems, integrity protects data and commands from unauthorized changes.[11] A second aspect of integrity requires that you must be able to detect if changes occur. The smart meter must be protected against concealed changes both physically and cyberwise. Since the smart meter is located at the customer site, the meter is vulnerable to tampering and vandalizing, and prevention from such physical attacks can be difficult. Customer gateways also must protect against undetected changes since they are conduits to critical customer equipment and systems.

10.6.2.3 Availability in AMI Systems   An attack on availability[11] makes resources inaccessible by authorized entities when they request them. The most important aspect to administer while ensuring availability in AMI systems is whether the data under question are affected by unavailability, and if affected, how time critical it is, whether in the scale of seconds, hours, or days. To overcome the unavailability problem, we have to provide creative approaches in routing the information between the smart meter, consumer gateway, and AMI communication networks. We also need to enable the smart meters to make local decisions. Detection methods for availability attacks include automated diagnostics and physical and cyber intrusion detection.

10.6.2.4 Nonrepudiation in AMI Systems Nonrepudiation ensures that the entities receiving the data do not subsequently deny receiving it, and if the entities did not receive the data, then they cannot subsequently state that they did receive it. Nonrepudiation in AMI systems[10] is important for all financial transactions. Also, the timeliness of response is as important as actually acting on a control command. Therefore, accurate timestamp information and continuous time synchronization across all AMI system components are crucial. Inaccurate timestamps and desynchronous messages lead to errors in customer information, billing for usage, and analysis for load and generation patterns by utility planners.

10.6.2.5 Authorization in AMI Systems Authorization in AMI systems[10] grants users and devices the right to access resources and perform specified actions. Lack of authorization will allow the AMI architecture to be vulnerable to attack from malicious elements that break into the network. As part of authorization, users and devices may be assigned roles, for example, that give them a set of privileges. By defining the scope of what an authenticated user or device can do, digital certificates can be used as an authorization mechanism.

10.6.3 Major Vulnerabilities in Current AMI Systems

The various major vulnerabilities in the current AMI systems are as follows:

10.6.3.1 Plaintext NAN Traffic Because of the rapid development and specification for the AMI systems, the implementation of decisions from vendors has been affecting the quality of security implementation in the system. Vendors may choose how to implement privacy and integrity control to protect the confidential data. In some cases, the vendors encrypt all traffic in the NAN; other vendors may decide not to use the encryption at some configuration level. For example, a product may have the capability of encryption but ship with a default setting of no encryption. This is a problem that can account for major security breaches.[12,13,14]

10.6.3.2 Bus Snooping   Embedded systems are used widely in peripheral devices such as radios that interface to measurement units. If the device has little or no physical protection, then a security risk may exist on the interfaces between the components in the embedded system. For example, the bus between the microcontroller and the radio is usually unencrypted which introduces vulnerability. The attacker can attach a bus sniffer onto the bus between the microcontroller and the radio to sniff packets.[15] The attacker is free to read and capture radio configuration information, cryptographic keys, network authentication credentials, and other sensitive information. Therefore, many radio chip manufacturers introduced cryptographic algorithms internally in hardware to prevent tampering with the chips and other components.

10.6.3.3 Improper Cryptography   Cryptography is easy to detect in the AMI infrastructure, but it is very difficult to detect if the cryptography is improperly configured. Improperly configured cryptography[12,13] could present a critical vulnerability to the infrastructure. The possible improper configuration of the cryptography could range from weak key derivation, improper reuse of key stream data, lack of replay, insecure cipher modes, weak integrity protection, insufficient key length, to cryptographically weak initialization vectors.

10.6.3.4 Direct Tampering   Tamper protection mechanisms are necessary to protect against malicious modification of the meter device installed in public or open areas.[16] The mechanisms could inevitably fail, but the tampering protection should delay attackers from damaging the integrity of data from the meter while notifying the utility company. The utility company should implement the ability to track the affected meter and alert law enforcement to capture the attackers.

When designing tamper protection mechanisms, the following list of features should be considered:

- Local tampering detection mechanism, which indicates any physical tampering with a meter
- Remote tampering detection mechanism, which notifies the head office that someone has been tampering with a meter
- Integrity-protecting mechanism, which protects and prevents modifying the sensitive information, such as security keys, meter configuration, and so on
- Repair authorization mechanism, which allows only authorized technicians or engineers from the utility company to repair the meter
- Physical lock mechanism, which prevents an unauthorized person from physically accessing the meter

10.6.3.5 Meter Authentication Weaknesses   The process of validating the credential passed between a meter and NAN device requires many steps. However, an attacker could impersonate a legitimate device and could gain information to undermine cryptographic protocols. Therefore, the process of meter authentication should be tested to make further improvements for defending authentication-related attacks during authentication exchanges.[12,17]

10.6.3.6 Denial-of-Service Threats   Denial of service is a common threat that prohibits access to the meter, and there are many conditions that trigger denial of service.[7] It is important to explore the possible DoS threats for meters.

10.6.3.7 Stored Key and Passwords   Because of the security requirements of meter devices, the manufacturers of AMI have included authentication, encryption, and integrity protections in the devices. Therefore, encryption keys, meter-derived keys, passwords, and other security-sensitive information are stored locally in the meter.[13] This presents an opportunity for hackers who compromise the meter to gain access to the NAN.

10.6.3.8 Cryptographic Key Distribution   Cryptography is supported in most radios and meters, but key management is a difficult problem.[12,18] For example, symmetric keys could be used in each meter, but if the attacker compromises any meter, the attacker will have access to the NAN or other meters and impersonate as a meter. Therefore, use of certificates, asymmetric keys, or public key infrastructure (PKI) is recommended. Two possible attacks from attackers are spoofing system update mechanisms to insert unauthorized certificates and allowing an attacker to decrypt and inject encrypted traffic.

10.7 Addressing Encryption and Key Management Needs of the Smart Grid Using Techniques Adapted from Sensor Networks

10.7.1 Data Encryption

To achieve the security goals stated in the previous sections, data encryption is essential. Some key security measures that have been developed for sensor networks can be adapted for addressing the security needs of the smart grid. A sensor network can be considered as a network of devices communicating using a short-range multihop communication infrastructure.

When the sender and the receiver use the same key for encryption, the mechanism is termed *symmetric key* cryptography. The sender uses the key to convert plaintext to ciphertext using the chosen encryption algorithm. The receiver recovers the plaintext from ciphertext using the same key and the corresponding decryption algorithm. On the other hand, *asymmetric key* or public key cryptography uses a unique (public, private) key pair for each communicating node. The public key of the node is used for encrypting data sent to the node. Since the private key is known only to the node, the data can be decrypted by the intended recipient only.

Symmetric key cryptography is computationally less intensive but does not scale well as each node requires a unique symmetric key with every other node in the network for successfully encrypting data between any two participating nodes. On the other hand, asymmetric key cryptography scales better but requires more computational resources.

If the devices are resource constrained, a symmetric key cryptosystem is more attractive, and most of the existing work in the literature is based on this methodology. Two types of ciphertext can be generated using symmetric key cryptography: stream cipher generated by encrypting the plaintext one bit at a time and block cipher generated by encrypting blocks of the plaintext at a time. The computational overhead of RC4 (stream cipher), IDEA and RC5 (block ciphers), and MD5 and SHA1 (one-way hash functions) have been evaluated in the work of Ganesan et al.[19] Different sensor platforms were used for testing these algorithms. It was shown that RC4 outperformed RC5 across all platforms. The hashing algorithms have an order of magnitude higher overhead compared to the symmetric key algorithms.

Several block ciphers were evaluated for their applicability in sensor networks in the work of Law et al.[20] The storage requirements and energy efficiency of the ciphers were also considered along with their security properties. The authors proposed Rjindel for applications with high security and energy efficiency requirements and MISTY1 for applications with both storage and energy efficiency needs.

Even though asymmetric key cryptography is not considered suitable for sensor networks, recent research has shown that it might be feasible with the proper choice of algorithms.[21,22] Asymmetric key cryptosystems are more scalable and resilient to node compromise. The challenge is to adapt the asymmetric key computation algorithms on the hardware design so that the computations can be supported by the resources available to the sensor nodes. Asymmetric key cryptosystems can be designed for sensor nodes with power consumption as low as 20 μW using optimized low-power techniques.[22] The future of public key encryption architectures for sensor networks looks promising with advances in sensor energy-harvesting techniques. Approaches based on elliptic curve cryptography (ECC) are also being investigated for sensor networks. TinyOS, the most widely used operating system for sensors, can be modified to support a public key infrastructure based on ECC.[23]

10.7.2 Key Establishment and Management

Of the different security measures, establishment of cryptographic keys is critical. Encryption as well as authentication mechanisms rely on them for their operation. The keys used by the cryptographic algorithms must be set up by the nodes before secure data exchange can take place. This process of establishing, distributing, and managing cryptographic keys is called *key management* and is one of the most challenging aspects of smart grid security design.

Security protocols rely on encryption mechanisms for ensuring data confidentiality. Also, for authentication purposes, the sender computes a message authentication code for each packet and appends to the message. Both the encryption algorithm and the message authentication code computation require cryptographic keys as inputs.[24] In a previous section, it was shown that symmetric key cryptography is preferred for sensor network applications. For large networks, it is extremely difficult to manage the creation and distribution of symmetric keys. Most symmetric key cryptosystems depend on a central authority for key creation and distribution. However, due to the lack of centralized control in some networks, this approach is not suitable.

For distributed networks, the simplest way to set up symmetric keys is to use a *networkwide key* for encryption and decryption purposes.[24] Hence, every node uses the same key for encryption and decryption. Although this approach does ensure data privacy and integrity, it is extremely vulnerable to node compromise since the sensor nodes are unattended for many applications. Even though this approach is simple to implement, it certainly is not an optimal solution.

The other extreme is to have *pairwise symmetric keys* preloaded for all sensor nodes in the network. However, the number of unique symmetric keys loaded in each sensor becomes unacceptably large as the size of the network increases. It has been proposed to use the sink as the key distribution center for setting up pairwise symmetric keys for the participating sensor nodes.[25] However, the sink becomes a single point of failure for the protocol. Also, it may lead to large communication overhead for sensors during the key exchange process.

In the work of Zhou and Fang,[26] it was shown that most recent approaches consider the key management problem for sensor networks as a two-step process. Prior to the deployment of the network, each sensor node is loaded with the initial keying material (*key predistribution phase*). This phase eliminates the dependence on the sink (or any other central node) for key distribution. The predistributed keying material depends on the memory resources of the sensor nodes and the resilience of the nodes to compromise. In other words, a node compromise should have an impact on a minimum number of nodes based on the information obtained from the predistributed material. Once the network is deployed, the nodes communicate with each other and establish either pairwise symmetric keys or asymmetric keys, based on the algorithms used (*key agreement phase*). Zhu et al.[27] showed that based on the communication pattern of the sensor nodes, a group key may also be established instead of pairwise keys.

The distribution of keying material can be probabilistic, deterministic, or hybrid.[28] In the probabilistic approach, each node is preloaded with a set of keys (key ring) randomly selected from a global key pool.[29,30] The neighboring nodes share at least one key with a certain probability depending on the size of the key ring, which in turn depends on the memory resources available. The challenge is to achieve a balance between the available resources and the desired key connectivity. Gong and Wheeler[31] presented deterministic approaches for key distribution that defined the global key pool and the key assignment to each node nonrandomly to increase the key connectivity between neighboring nodes. Instead of uniformly distributing the keying material across the entire network, a location-based key material distribution system can be used to optimize one-hop key connectivity.[32]

Most of the existing sensor security solutions rely on a key predistribution mechanism to alleviate the problem of key distribution and management. Others rely on the sink for key distribution. Both of these approaches are not optimal, and the design of key management schemes for sensor networks is still an open research problem.

10.7.3 Link-Layer Security Frameworks

A few years ago, the focus of sensor network research was key management. Another area of interest was secure routing. However, recent work has been in the area of link-layer security frameworks in the quest for a more general solution that can be used for different applications and situations. Link-layer security works with sensor network features such as in-network processing and data aggregation. These features enable the sensed data to be processed and aggregated at each intermediate node so that unnecessary transmissions can be avoided. Note that the energy used in processing is less by several orders of magnitude compared to the energy in every bit of information that is transmitted and received by sensors. Also, end-to-end security solutions can be subjected to certain DoS attacks, which can be prevented by link-layer security architectures that can detect malicious packets injected in the network at an early stage. Several link-layer approaches exist in the literature for addressing the security needs of sensor networks and provide another tool for smart grid security.[24,33–39]

10.8 Conclusions and Outlook

The distributed, changing, and physically exposed nature of the smart grid makes it more susceptible to cyber attacks than many existing networks. A security analysis of the smart grid communication architecture indicated several likely attack methods. Security solutions from traditional networks and from sensor networks can be adapted to the smart grid. An essential smart grid security feature is the ability to detect compromised nodes and for nodes to be able to send notification if they are attacked. In addition, for the smart grid to maintain effective encryption of private data and prevent attacks, an effective key management system must be used. This is an area of active, ongoing research.

References

1. German Federal Office for Information Security (2011). *Protection Profile for the Gateway of a Smart Metering System*, v01.01.01 final draft. Federal Office for Information Security, Bonn, Germany.
2. National Association of State Energy Officials (NASEO) (December 2010). *Smart Grid and Cyber Security for Energy Assurance—Planning Elements for Consideration in States' Energy Assurance Plans*. NASEO, Arlington, VA.
3. National Institute of Standards and Technology (NIST) (August 2010). *NISTIR 7628 Guidelines for Smart Grid Cyber Security, Introduction and Volumes 1–3*. Cyber Security Coordination Task Group, Advanced Security Acceleration Project Smart Grid, NIST, Gaithersburg, MD.
4. C. Bennett, B. Brown, B. Singletary, D. Highfill, D. Houseman, F. Cleveland, H. Lipson, J. Ivers, J. Gooding, J. McDonald, N. Greenfield, and S. Li (December 2008). *AMI System Security Requirements*, Utility Communication Architecture International User Group (UCAIUG), Raleigh, NC.
5. H. Khurana, M. Hadley, N. Lu, and D. A. Frincke (2010). Smart-grid security issues. *IEEE Security and Privacy*, doi: 10.1109/MSP.2010.49, pp. 81–85.
6. D. C. Schleher (July 1999). *Electronic Warfare in the Information Age*. Artech House, Norwood, MA.

7. Z. Lu, X. Lu, W. Wang, and C. Wang (October 2010). Review and evaluation of security threats on the communication networks in the smart grid. *Military Communications Conference, 2010—Milcom 2010*, doi: 10.1109/MILCOM.2010.5679551, pp. 1830–1835.

8. C. Karlof and D. Wagner. Secure routing in wireless sensor networks: attacks and countermeasures. *First IEEE International Workshop on Sensor Network Protocols and Applications*, Anchorage, AK (May 2003).

9. Problems with Smart Grid. eHow.com. http://www.ehow.com/info_8072577_problems-smart-grid.html#ixzz1ImKhD3c8

10. Open SG User Group. *AMI Security Specification v_2.01*, Nashville, TN. http://osgug.ucaiug.org/utilisec/amisec/default.aspx

11. C. Bennett and D. Highfill (November 2008). Networking AMI smart meters. *Energy 2030 Conference, 2008. ENERGY 2008*. IEEE, New York, pp. 1–8.

12. M. Carpenter, T. Goodspeed, B. Singletary, E. Skoudis, and J. Wright (January 5, 2009). *Advanced Metering Infrastructure Attack Methodology*. http://www.inguardians.com/pubs/articles.html

13. F. M. Cleveland (July 2008). Cyber security issues for advanced metering infrastructure (AMI). *Power and Energy Society General Meeting—Conversion and Delivery of Electrical Energy in the 21st Century, 2008*, IEEE, New York, pp. 1–5.

14. M. Theoharidou, G. Marias, S. Dritsas, and D. Gritzalis (2006). The ambient intelligence paradigm. A review of security and privacy strategies in leading economies. *2nd IET International Conference on Intelligent Environments. IE 06*, vol. 2, pp. 213–219.

15. R. Chaki (October 2010). Intrusion detection: ad-hoc networks to ambient intelligence framework. *International Conference on 2010 Computer Information Systems and Industrial Management Applications (CISIM)*, pp. 7–12.

16. A. Hahn (September 2010). Smart grid architecture risk optimization through vulnerability scoring. *2010 IEEE Conference on Innovative Technologies for an Efficient and Reliable Electricity Supply (CITRES)*, pp. 36–41.

17. C. Bennett and S. B. Wicker (July 2010). Decreased time delay and security enhancement recommendations for AMI smart meter networks. In *Innovative Smart Grid Technologies (ISGT), 2010*, pp. 1–6.

18. J. Kim, S. Ahn, Y. Kim, K. Lee, and S. Kim (June 2010). Sensor network-based AMI network security. *2010 IEEE PES Transmission and Distribution Conference and Exposition*, pp. 1–5.

19. P. Ganesan et al. (September 2003). Analyzing and modeling encryption overhead for sensor network nodes. *Proceedings of 2nd International Conference on Wireless Sensor Network Applications*, pp. 151–159.

20. Y. W. Law, J. Doumen, and P. Hartel (November 2006). Survey and benchmark of block ciphers for wireless sensor networks. *ACM Transactions on Sensor Networks*, 2, 65–93.

21. R. Watro et al. (November 2004). TinyPK: securing sensor networks with public key technology. *Proceedings of 2nd ACM Workshop on Security of Ad Hoc and Sensor Networks (SASN'04)*, Washington, DC.

22. G. Gaubatz, J. Kaps, and B. Sunar (October 2005). *Public Key Cryptography in Sensor Networks—Revisited. Lecture Notes in Computer Science—Security in Ad-Hoc and Sensor Networks.* Springer, New York.

23. D. J. Malan et al. (October 2004). A public-key infrastructure for key distribution in TinyOS based on elliptic curve cryptography. *Proceedings of 1st IEEE International Conference on Sensor and Ad Hoc Communication Networks (SECON'04)*, Santa Clara, CA.

24. C. Karlof et al. (November 2004). TinySec: a link layer security architecture for wireless sensor networks. *Proceedings of 2nd International Conference on Embedded Networked Sensor Systems (SenSys'04)*, pp. 162–175.

25. A. Perrig et al. (2002). SPINS: Security Protocols for Sensor Networks. *ACM Wireless Networks*, 8(5), 521–534.

26. Y. Zhou and Y. Fang (2008). Securing wireless sensor networks: a survey. *IEEE Communications Surveys and Tutorials*, 10(3), 6–28.

27. S. Zhu et al. (October 2003). LEAP: efficient security mechanism for large scale distributed sensor networks. *Proceedings of 10th ACM Conference on Computer and Communications Security (CCS'03)*, pp. 62–72.

28. S. Camtepe et al. (2008). *Key Management in Wireless Sensor Networks. Wireless Sensor Network Security.* J. Lopez and J. Zhou (Eds.). IOS Press, Amsterdam, the Netherlands.

29. H. Chan et al. (June 2006). Random key predistribution schemes for sensor networks. *IEEE International Conference on Communication*, pp. 2262–2267.

30. L. Eschenauer and V. Gligor (November 2002). A key management scheme for distributed sensor networks. *Proceedings of 9th ACM Conference on Computer and Communications Security (CCS'02)*, pp. 41–47.

31. L. Gong and D.J. Wheeler (1990). A matrix key distribution scheme. *Journal of Cryptology*, 2(1), 51–59.

32. D. Liu and P. Ning (October 2003). Location-based pairwise key establishments for relatively static sensor networks. *Proceedings of 2003 ACM Workshop Security of Ad Hoc and Sensor Networks (SASN'03)*, Fairfax, VA USA.

33. Q. Xue and A. Ganz (October 2009). Runtime security composition for sensor networks (secure sense). *IEEE 58th Vehicular Technology Conference (VTC'03)*, pp. 2976–2980.

34. N. Sastry and D. Wagner (October 2004). Security considerations for IEEE 802.15.4 networks. *ACM Workshop on Wireless Security (Wise'04)*, pp. 32–42.

35. T. Li, H. Wu, X. Wang, and F. Bao (May 2005). SenSec: sensor security framework for TinyOS. *Proceedings of 2nd International Workshop on Networked Sensing Systems (INSS'05)*, San Diego, CA.

36. A. D. Wood et al. (October 2006). SIGF: a family of configurable, secure routing protocols for wireless sensor networks. *Proceedings of Fourth ACM Workshop on Security of Ad Hoc and Sensor Networks (SASN'06)*.

37. M. Luk, G. Mezzour, A. Perrig, and V. Gligor (April 2007). MiniSec: a secure sensor network communication architecture. *IEEE International Conference on Information Processing in Sensor Networks (IPSN'07),* Cambridge, MA.

38. P. Osanacek (2009). *Towards Security Issues in ZigBee Architecture. Lecture Notes in Computer Science—Human Interface and Management of Information, Designing Information Environments.* Springer, New York.

39. M. Healy, T. Newe, and E. Lewis (2009). Security for wireless sensor networks: a review. *IEEE Sensors Applications Symposium,* New Orleans, LA.

Recommended Reading

A. Agah and S. Das (2007). Preventing DoS attacks in wireless sensor networks: a repeated game theory approach. *International Journal of Network Security,* 5(2), 145–153.

E. Cayirci and C. Rong (2009). *Security in Wireless Ad Hoc and Sensor Networks.* Wiley, West Sussex, UK.

H. Chan and A. Perrig (2003). Security and privacy in sensor networks. *IEEE Computer Magazine,* 36(10), 103–105.

B. Deb, S. Bhatnagar, and B. Nath (2003). Information assurance in sensor networks. *Proceedings of 2nd ACM International Conference on Wireless Sensor Networks and Applications,* pp. 160–168.

J. Deng, R. Han, and S. Mishra (2002). *INSENS: Intrusion Tolerant Routing in Wireless Sensor Networks.* Technical Report CU-CS-939-02. Department of Computer Science, University of Colorado at Boulder.

J. R. Douceur (2002). The Sybil attack. *Proceedings of 1st International Workshop on Peer-to-Peer Systems (IPTPS'02),* pp. 251–260.

D. Han, J. Zhang, Y. Zhang, and W. Gu (2010). Convergence of sensor networks/Internet of things and power grid information network at aggregation layer. *2010 International Conference on Power System Technology (POWERCON),* doi: 10.1109/POWERCON.2010.5666553, pp. 1–6.

C. Hartung, J. Balasalle, and R. Han (2005). *Node Compromise in Sensor Networks: The Need for Secure Systems.* Technical Report CU-CS-990-05. Department of Computer Science, University of Colorado at Boulder.

F. Hu and N. K. Sharma (2005). Security considerations in ad hoc sensor networks. *Elsevier Ad hoc Networks,* 3(1), 69–89.

Y. C. Hu, A. Perrig, and D. B. Johnson (2003). Packet leashes: a defense against wormhole attacks in wireless ad hoc networks. *Proceedings of INFOCOM,* pp. 1976–1986.

Institute for Electrical and Electronics Engineers (September 10, 2011). *IEEE Standard 2030™—2011 Guide for Smart Grid Interoperability of Energy Technology and Information Technology Operation with the Electric Power System (EPS), End-Use Applications, and Loads.* IEEE, New York.

C. Karlof and D. Wagner (2003). Secure routing in wireless sensor networks: attacks and countermeasures. *Ad Hoc and Sensor Networks*, 293–315.

R. A. Kisner et al. (2010). Cybersecurity through Real-Time Distributed Control Systems, Oak Ridge National Lab, ORNL/TM-2010/30. Oak Ridge National Lab, Oak Ridge, TN.

J. Lopez and J. Zhou (2008). *Wireless Sensor Network Security*. IOS Press, Amsterdam, Netherlands.

O. Komerling and M. G. Kuhn (May 1999). Design principles for tamper resistant smartcard processors. Paper presented at USENIX Workshop on Smartcard Technology, Chicago.

M. Mohi et al. (2009). A Bayesian game approach for preventing DoS attacks in wireless sensor networks. *Proceedings of the 2009 WRI International Conference on Communications and Mobile Computing*, Vol. 3, pp. 507–511.

Moog Crossbow (2010). Crossbow Mica2 Motes. http://www.xbow.com

J. Newsome et al. (2004). The Sybil attack in sensor networks: analysis and defenses. *Proceedings of 3rd International Symposium on Information Processing in Sensor Networks*. ACM Press, New York.

B. Parno, A. Perrig, and V. Gligor (2005). Distributed detection of node replication attacks in sensor networks. *Proceedings of IEEE Symposium on Security and Privacy*, Oakland, CA.

A. Perrig, J. Stankovic, and D. Wagner (2004). Security in wireless sensor networks. *Communications of ACM*, 47(6), 53–57.

G. Pottie and W. Kaiser (2000). Wireless integrated network sensors. *Communications of the ACM*, 43(5), 51–58.

E. Shi and A. Perrig (2004). Designing secure sensor networks. *IEEE Wireless Communications Magazine*, 11(6), 38–43.

H. Song, L. Xie, S. Zhu, and G. Cao (2007). Sensor node compromise detection: the location perspective. *Proceedings of International Conference on Wireless Communication and Mobile Computing*, pp. 242–247.

M. Tubaishat, J. Yin, B. Panja, and S. Madria (2004). A secure hierarchical model for sensor network. *ACM SIGMOD Record*, 33, 7–13.

J. Undercoffer et al. (2002). Security for sensor networks. Paper presented at *CADIP Research Symposium*, Baltimore.

J. Walters et al. (2006). Wireless sensor network security: a survey. In Y. Xiao (Ed.) *Security in Distributed, Grid and Pervasive Computing*, pp. 367–410. CRC Press, Boca Raton, FL.

Y. Wang, G. Attebury, and B. Ramamurthy (2006). A survey of security issues in wireless sensor networks. *IEEE Communication Surveys and Tutorials*, 8(1), 2–23.

A. D. Wood and J. A. Stankovic (2002). Denial of service in sensor networks. *Computer*, 35(10), 54–62.

J. Yick et al. (2008). Wireless sensor network survey. *Elsevier Computer Networks*, 52(12), 2292–2330.

List of Acronyms

A2A: Application to Application
AAA: authentication, authorization, and accounting
ACSE: association control service element
ACSI: abstract communication service interface
ADC: analog-to-digital converter
AES: Advanced Encryption Standard
AGA: American Gas Association
AH: Authentication Header
AMI: advanced metering infrastructure
AMR: advanced meter reading
ANSI: American National Standards Institute
AP: access point
APCO: Association of Public-Safety Communications Officials
APDU: Application Protocol Data Unit
API: application program interface
ARM: advanced RISC machine
ARP: Address Resolution Protocol
ASDU: application service data unit
ASN.1: Abstract Syntax Notation One
AWGN: additive white Gaussian noise
BAS: building automation system
B2B: Business to Business

BES: bulk electric system
B2G: Building-to-Grid (a)
BMS: building management system
B&P: Business and Policy
BPL: broadband over power line
BS: base station
BSS: blind source separation
CA: certificate authority
CDC: Common Data Class
CDPSM: Common Distribution Power System Model
CHAP/PAP: Challenge Handshake Authentication Protocol/ Password Authentication Protocol
CHP: combined heat and power
CIGRE: International Council on Large Electronic Systems
CIM: Common Information Model
CIMug: CIM Users Group
CIP: Critical Infrastructure Protection
CIS: Component Interface Specification
CLS: controllable local system
CMDA: code division multiple access
COSEM: Companion Specification for the Energy Metering
CPC: chaining block cipher
CPE: customer premises equipment
CPP: critical peak pricing
CPSM: Common Power System Model
CPU: central processing unit
CR: cognitive radio
CRL: Certificate Revocation List
CSCTG: Cyber Security Coordination Task Group
CSMA: carrier sense multiple access
CT: current transformer
DA: distribution automation
DAC: digital-to-analog converter
DAP: day-ahead pricing
DCCP: Datagram Congestion Control Protocol
DCS: distributed control system
DDoS: distributed DoS
DER: Distributed Energy Resources

3DES: Triple Data Encryption Algorithm

DEWG: domain expert working group

DLC: distribution line carrier

DLMS: Distribution Line Message Specification

DMS: distribution management system

DNP3: Distributed Network Protocol

DNS: Domain Name System

DoF: degrees of freedom

DoS: denial of service (hyphen if adj)

DP: development platform

DPO: digital phosphor oscilloscope

DR: demand response

DSP: digital signal processor

DSS: digital signature standard

DTLS: Datagram Transport Layer Security

DVFS: dynamic voltage and frequency scaling

EAI: Enterprise Application Integration

EAP: Extensible Authentication Protocol

ECC: elliptic curve cryptography

EDIFACT: Electronic Data Interchange for Administration, Commerce, and Transport

EMS: energy management system

ENTSO-E: European Network of Transmission System Operators for Electricity

EPRI: Electric Power Research Institute

EPSEM: Extended protocol specification for electronic metering

ERCOT: Electric Reliability Council of Texas

ESB: enterprise service bus

ESP: Encapsulated Security Payload

FAN: field-area network

FEP: front-end processor

FFT: fast Fourier transform

FHSS: frequency-hopping spread spectrum

FIPS: Federal Information Processing Standard

FPGA: field-programmable gate array

FSK: frequency shift keying

FMSC: finite-state Markov chain

GDOI: Group Domain of Interpretation

GES: generic eventing and subscription
GID: Generic Interface Definition
GMAC: Galois Message Authentication Code
GOOSE: Generic Object Oriented Substation Event
GPRS: general packet radio services
GSM: Global System for Mobile Communications
GSSE: Generic Substation Status Event
GWAC: GridWise Architecture Council
HAL: hardware abstraction layer
HAN: home-area network
HCB: hybrid cloud broker
HCBC: hash-CBC
HDLC: High-Level Data Link Control
HiperLAN: High Performance Radio LAN
H2G: Home-to-Grid (a)
HMAC: hash message authentication code
HMI: human-machine interface
HSDA: High-speed data access
HTTP: Hypertext Transfer Protocol
IACS: industrial automation and control system
IBR: inclining block rate
IALM: inexact augmented Lagrange multiplier
I/C: interruptible/curtailable
ICA: independent component analysis
ICMP: Internet Message Control Protocol
ICS: industrial control system
ICS: Industrial Control System Security (of NIST)
IDS: intrusion detection system
IEC: International Electrotechnical Commission
IED: intelligent electrical device
IEEE: Institute of Electrical and Electronics Engineers
IETF: Internet Engineering Task Force
I2G: Industrial-to-Grid (a)
i.i.d.: independent and identically distributed
IPSec: Internet Protocol Security
IPv4: Internet Protocol Version 4
IRM: Interface Reference Model
ISA: International Society of Automation

ISC: Industrial Control Systems

ISC-CERT: Industrial Control Systems Cyber Emergency Response Team

ISDN: Integrated Services Digital Network

ISO: International Organization for Standardization

ITU: International Telecommunication Union

IV: initialization vector

JMS: Java Messaging Service

KPCA: kernel PCA

LAN: local-area network

LCE: loosely coupled event

LD: logical device

LLC: logical link control

LMDS: local multipoint distribution service

LMN: local metrological network

LMR: land mobile radio

LMVU: landmark maximum variance unfolding

LN: logical node

LTC: load tap changer

MAC: message authentication code

MCM: multicarrier modulation

MAN: metropolitan-area network

MDA: Model Driven Architecture

MDI: meter data integration

MIMO: multiple input multiple output

MDMS: meter data management system

MIB: Management Information Base

MMS: Manufacturing Message Specification

MOM: message-oriented middleware

MPSL-VPN: Multi-Protocol Label Switching-Virtual Private Network (MPLS-VPN)

MSPS: mega-samples per second

MV: medium-voltage (a)

MVU: maximum variance unfolding

NAN: neighborhood-area network

NASEO: National Association of State Energy Officials

NERC: North American Reliability Corporation

NetAPT: Network Access Policy Tool

NIPP: National Infrastructure Protection Plan
NOSR: no optimal stopping rule
NP: nondeterministic polynomial
NIST: National Institute of Standards and Technology
NRECA: National Rural Electric Cooperative Association
NSM: network and system management
OBIS: object identification system
OCSP: Online Certificate Status Protocol
OFDMA: orthogonal frequency-division multiple access
OMG: Open Management Group
ORBIT: Open Access Research Testbed for Next-Generation Wireless
        Networks
OS: operating system
OSI: Open System Interconnection
OSR: optimal stopping rule
OSSTMM: Open Source Security Testing Methodology Manual (ch8)
PAD: packet assembler-disassembler
PAN: personal area network
PAR: peak-to-average ratio
PCA: principal component analysis
PCIe: Peripheral Component Interconnect Express
PDU: protocol data unit
PGP/GnuPG: pretty good privacy/Gnu Privacy Guard
PHEV: plug-in hybrid electric vehicle
PIM: Platform Independent Model
PKI: public key infrastructure
PLC: programmable logic controller
PN: pseudorandom noise
PSD: positive semidefinite
PSM: Platform Specific Model
PSTN: public switched telephone network
QoS: quality of service
QPSK: quadrature phase shift keying
RCB: radio control board
RDF: Resource Description Framework
RFC: Request for Comments
RISC: reduced instruction set computing
RSA: Rivest–Shamir–Adleman

RTP: real-time pricing
RTU: remote terminal unit
SAML: Security Assertion Markup Language
SAN: storage area network
SAS: Substation Automation System
SB: site broker
SCADA: supervisory control and data acquisition
SCL: Substation Configuration Language
SCSM: Specific Communication Service Mapping
SCTP: Stream Control Transmission Protocol
SDP: semidefinite programming
SDR: software-defined radio
SFF: small form factor
S-FSK: spread frequency shift keying
SG3: Smart Grid Strategic Group
SGAM: Smart Grid Architectural Model
SGCG: Smart Grid Coordination Group
SGiP: Smart Grid Interoperability Panel
SHA-1: secure hash algorithm
SIA: Seamless Integration Architecture
SIDM: system interfaces for distribution management
SIR: signal-to-interference ratio
SLA: service-level agreement
SLO: service-level objective
SM: smart meter
S/MIME: secure/multipurpose Internet mail extensions
SMV: sample measured value
SNMP: Simple Network Management Protocol
SNR: signal-to-noise ratio
SNTP: Simple Network Time Protocol
SOA: service-oriented architecture
SOAP: Simple Object Access Protocol
SoC: System-on-Chip
SOHO: small office/home office
SP: Special Publication
SRTP: Secure Real-Time Transport Protocol (SRTP)
SS-AW: Spread spectrum adaptive wideband
sSCADA: secure SCADA

SS-FFH: Spread spectrum–fast frequency hopping
SSH: Secure Shell
SSL: Secure Sockets Layer
SSPP: Serial SCADA Protection Protocol
SV: Sample Value
SVD: singular value decomposition
SVM: support vector machine
TC: Technical Committee
TCP/IP: Transmission Control Protocol/Internet Protocol
TCIPG: Trustworthy Cyber Infrastructure for the Power Grid
T&D: Transmission and Distribution
TLS: Transport Layer Security
TOU: time-of-use (a)
TPDU: Transport Protocol Data Unit
TR: Technical Report
TSDA: time series data access
TSEL: transport selector
TTP: trusted third party
UCAIug: Utility Communication Architecture International User Group
UDDI: Universal Description, Discovery, and Integration
UDP: User Datagram Protocol
UML: Unified Modeling Language
URI: Uniform Resource Identifier
USRP2: Universal Software Radio Peripheral 2
UWB: ultra-wideband
VLAN: virtual local-area network
VM: virtual machine
VoIP: Voice over Internet Protocol
VPN: virtual private network
VT: voltage transformer
WAM: wide-area measurement system
WAN: wide-area network
WARP: Wireless Open-Access Research Platform
WBX: wide bandwidth transceiver
W3C: World Wide Web Consortium
WEP/WAP: wired equivalent privacy
WG: working group

WOL: wake-on-LAN
WRAN: wireless regional-area network
WSDL: Web Services Description Language
WSN: wireless sensor network
WS-Security: web services security
WS-Trust: Web Services Trust
XML: eXtensible Markup Language

Index

Milton Keynes UK
Ingram Content Group UK Ltd.
UKHW031142141024
449569UK00024B/1136